PLANTS, MAN AND THE LAND
IN THE VILCANOTA VALLEY OF PERU

BIOGEOGRAPHICA

VOLUME VI

DR. W. JUNK B.V., Publishers, The Hague 1975

PLANTS, MAN AND THE LAND IN THE VILCANOTA VALLEY OF PERU

by

DANIEL W. GADE

DR. W. JUNK B.V., Publishers, The Hague 1975

ISBN-13:978-94-010-1963-7 e-ISBN-13:978-94-010-1961-3
DOI: 10.1007/978-94-010-1961-3

Softcover reprint of the hardcover 1st edition 1975

Cover design Max Velthuijs, The Hague

CONTENTS

PREFACE

Man's symbiosis with plants is the most fundamental material fact of human life on the earth. Geographers, as well as botanists, anthropologists and other scientists, have long been interested in this aspect of the man-nature theme. In American geography, CARL O. SAUER emphasized a temporal as well as spatial perspective in the cultural understanding of man's relationship to biological phenomena. His researches and those of his associates in the 'Berkeley school' showed that the most fruitful possibilities for implementing this approach are in non-industrial societies which have direct and pervasive links between plants and man (GADE, 1975). The study that follows is a geography of plant resources in an important Andean valley having great environmental diversity and a cultural constant, in so far as a non-literate, Quechua-speaking peasantry dominates throughout the zone. My basic objective has been to understand the present use of plants, cultivated and wild, as they have varied from place to place and through time.

Primary and secondary documents and local informants were important sources of historical information. Most of the contemporary data in this study were derived from over 20 months of empirical observations of the day-to-day existence of farming folk in their fields, homes and markets. The great natural beauty of the Vilcanota depression is matched only by the stark poverty which has been the lot of the majority of people who live there. My purpose was not, however, to explore the human potential of the population or the possibilities for socio-economic development in the area. Hopefully it will contribute to an understanding and appreciation of a highly traditional region of South America, and the basis of livelihood of a people whom many technicians and planners feel compelled to change.

The content of this book goes beyond the realm of most treaties on ethnobotany, and is better described as 'cultural biogeography.' Chapter 1 is based on the assumption that plant-man relationships must be seen in the larger context of place, time, culture and livelihood. Chapter 2 organizes the use of plants by the Vilcanota peasants into their temporal, ecological and spatial dimensions. Chapter 3 discusses the useful and conspicuous plants of the valley within a taxonomic framework, in this way presenting what is important about a species without those presuppositions involved in categorizing plants solely according to use and/or distribution. Chapter 4 briefly conceptualizes the dominant themes.

Most of the material in this work resulted from field research in Highland Peru from June 1963 to December 1964, supported by the Foreign Field Research Program and administered by the Division of Earth Sciences, National Academy of Sciences-National Research Council, under sponsorship of the Geography Branch, Office of Naval Research, Contract No. Nonr-2300(09), Subcontract No. 63. Subsequent field studies in Peru between January and May 1968 (sponsored by the University of Vermont) and from May to August 1970 (funded by the American Council of Learned Societies and the Social Science Research Council) have corroborated, expanded or amended some of the original field data. Since these field periods, some shifts have occurred in the land-holding arrangements of the study area, but these have not significantly altered the cultural biogeography.

I am grateful to many individuals in Peru and the United States for their assistance in the formulation, execution and completion of this work. Several hundred Quechua peasants tolerantly allowed me to observe them and their plants without fully understanding my motives. Among the many *cuzqueños* who gave of their time and knowledge, I would like to mention Dr. CÉSAR VARGAS CALDERÓN, Dr. MARIO ESCOBAR MOSCOSO and Ing. VICTOR OCHOA OLIART. Valuable stimulation and useful suggestions came from a creative group of scholars who were associated with the University of Wisconsin in Madison in the early and middle 1960's, most notably Drs. WILLIAM M. DENEVAN, FREDERICK J. SIMOONS, HENRY S. STERLING, HUGH H. ILTIS, CLARENCE W. OLMSTEAD and JONATHAN D. SAUER. MARY SCOTT KILLGORE GADE made a major contribution by her geographical understanding and editorial and cartographic skills. Of course I alone am responsible for errors or shortcomings in the work.

DANIEL W. GADE

Burlington, Vermont, 12 April 1975

1. MAN AND THE LAND: CHARACTER OF THE AREA

The land of Peru is conventionally divided into three discrete regions because of their highly contrasting physical and human environments. To the west is a desert strip, the Coast or *Costa*, where the rainfall is everywhere less than 250 millimeters, but where a humid, cloudy atmosphere prevails due to the presence of the cold Peru Current offshore. The region is cut by transverse valleys which are agriculturally productive where irrigation has been possible. Today in this region about half a million hectares are in crops, mostly in large farm units on which irrigated sugar cane and cotton are produced for export and rice for internal consumption. Ninety percent of Peru's agricultural exports are from the Coast, although twice as much land is under cultivation in the Sierra.

The Highlands or Sierra refers to a complex landscape of hills, snow-covered mountains, deep and narrow gorges that widen out into broad valleys toward the east. Most of the Sierra is sub-humid with a distinct seasonality of precipitation. Although large areas in the Highlands may be completely empty of people, virtually all land with any agricultural possibilities is exploited, and the man-land density is about one person to 1/4 hectare of arable land. Human population is especially dense in temperate valleys where irrigated agriculture is possible. Most products are for subsistence purposes or are sold in local markets; however, some – such as wheat and potatoes – which are not grown much on the Coast, reach the rest of the country.

The eastern Lowlands or Montaña includes the lower eastern slopes of the Andes and the Amazon flatlands and is characterized by uniformly high monthly temperatures and generally more than 1,500 millimeters of rainfall per year. The Montaña is sparingly farmed and, in contrast to the Sierra, usually by migratory slash and burn agricultural methods. The native forest peoples in this vast region have rapidly declined in number, and contact with former mountain dwellers has become the rule rather than the exception.

The Land of Peru in Cultural-Historical Perspective

In these three broad settings, culture developed at different times and with varying degrees of sophistication. Although most of our prehistoric evidence of culture is from the Coast because of the differential preservation of materials, it is becoming increasingly clear that the three regions were bound together in trade for at least 2,000 years prior to the Spanish Conquest. Agriculture as the basis of subsistence appeared on the Coast around the beginning of important maize cultivation (*c.* 1200 B.C.), but the period of experimentation with domesticated crops extended at least 1,500 years before that (TOWLE, 1961). The idea of plant cultivation was very probably introduced to western South America from elsewhere. Prior to about 2500 B.C. a hunting and gathering economy may be postulated for the Sierra; after this time a gradual spread of agriculture eventually resulted in sedentary societies that made pottery. The degree to which civilization developed in the Sierra before the time of Christ is still not clear. We know that the Coast was the scene of remarkable achievements in arts, technology and political organization, but the same cannot be said with assurance for the Sierra before A.D. 500. Little careful or systematic excavation has been carried out in the Highlands and pre-Inca cultures over much of this rugged region are known to us mainly through surface ruins.

Intelligent speculation (SAUER, 1952:40-49) and accruing archeological evidence (LATHROP, 1965:796-8; PICKERSGILL, 1969), suggest that from the Eastern Lowlands came major contributions to the overall cultural development and achievements of the entire Andean realm. One of the first to express this idea was TELLO (1930) who believed that the first high culture originated in the Highlands by people who had migrated from the eastern forest with some knowledge of agriculture. The cultural development was diffused to the Coast where the coastal dwellers replaced their fishing economy with plant cultivation from the Highlands to the east. It would appear that currents of diffusion from both the east as well as from the north occurred, but, based on our present knowledge, it is not clear which one predominated.

A thousand years after the time of Christ a group commonly referred to as the Inca arose and gradually extended hegemony over many peoples. The Inca Empire, known in Quechua as *Tawantinsuyo*, occupied 45 percent of the territory of present-day Peru and extended north to northern Ecuador and south to central Chile (Fig. 1). Four sections comprised this organized territory, corresponding more or less to different physical environments. *Collasuyu,* the southern division, was the largest in area and included the Titicaca Basin and most of present-day Bolivia, northwest Argentina and the northern half of Chile. *Cuntisuyu* covered much of the semiarid and arid central highlands west of Cuzco, reaching the coast. *Chinchasuyu* was the northern part of the realm, and extended along the coast, west of *Cuntisuyu*, to about the site now occupied by Ica. *Antisuyu* was the eastern division taking in the middle and lower eastern Andean slopes; its north-south extent was not well defined. Cuzco ('the navel') was at the heart of the Empire, the nerve center from which radiated the orders holding all parts together. The accurate population records kept by the Inca were destroyed when the Spaniards came; population estimates for the Inca Empire vary enormously from the thirty million of DOBYNS (1966:415) to the three million of KROEBER (1939:166).

The material foundation of the Inca Empire was a highly productive and sustained agriculture, but one based on plants that were first cultivated and animals that were first domesticated many centuries before the Incas ever appeared on the Andean scene. The Inca Empire was, in fact, but a moment in the history of the Andes and most of the accomplishments were made by generations of obscure folk

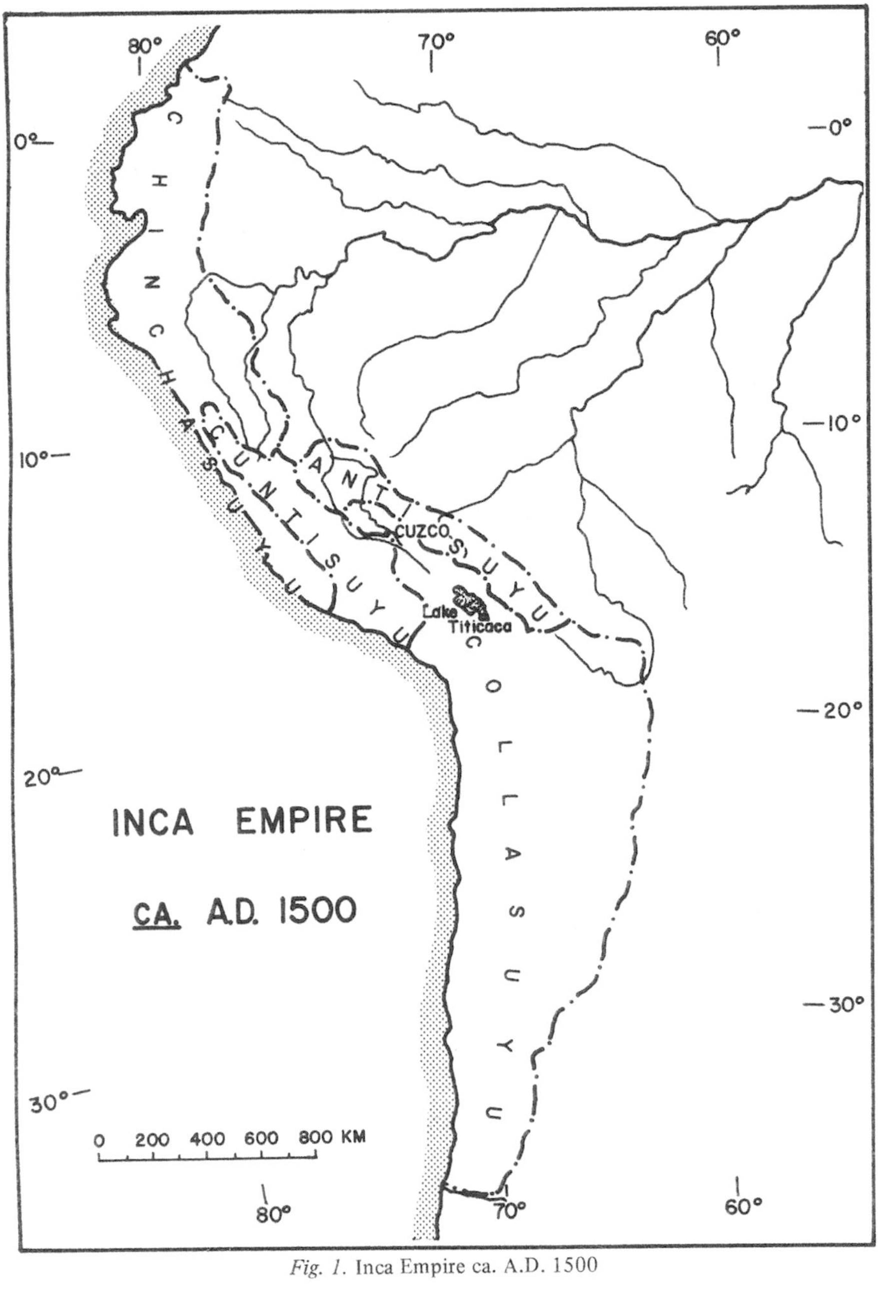

Fig. 1. Inca Empire ca. A.D. 1500

who preceded in time this political amalgamation. Because of the relatively humid conditions that occur in much of the Highlands, very little preserved plant material from the pre-Inca or Inca periods has been found in the Sierra. The oldest materials discovered so far are potato and quinoa remains at a Bolivian site near Lake Titicaca dated at *c.* 500 B.C. (TOWLE, 1961:8). Most archeological evidence of agriculture and plant use in this mountain region is inferential in nature ranging from the presence of round storage structures for crops to mortars and pestles used to grind plant foods.

TROLL (1943:93-128) asserted that the southern Peruvian Highlands had several important natural advantages that made it, rather than northern Peru and Ecuador or Chile, the center of the Inca Empire. These were the below freezing temperatures, a condition necessary for the preparation of dehydrated potato which was a critical source of food in unproductive periods; the presence of *ichu* grass, the natural food of llamas, the only beast of burden; and longer days which permitted a higher elevational limit for the growth of crops. Yet no matter how much each of these three factors may have facilitated the establishment of the Empire, without the remarkable organizational ability of the Incas themselves no such broad political unit would have been formed, whatever the nature of the physical environment and resource base.

In the heart of the former Inca Empire and near its probable center or origin lies the Vilcanota Valley, throughout history an important zone of human settlement in the southern Peruvian Highlands. Located in the Department of Cuzco, the Vilcanota Valley was selected as the area for this study because of its long occupance, wide range of natural environments, and accessibility for lengthy field research. The southern limit of the study area is the head of drainage and the northern limit is set at the confluence of the Yanatile River. Although the Urubamba (lower part of the Vilcanota) River continues several hundred kilometers north of this point to join the Ucayali, the Yanatile juncture represents the northern limit of traditional Quechua settlement and the point at which the wooded savanna landscape gives way to a tropical rainforest. Although this study locale is only one of a host of eastern Andean depressions that become part of the Amazon drainage, many of the findings are applicable to other eastern Andean valleys as well. Some of the generalizations are valid for the whole Peruvian Highlands, as well as for the entire Andean culture region. Still other statements of a conceptual nature are true for most peasant societies, if not throughout the world, at least in mainland Latin America.

The Natural Setting of the Vilcanota Valley

The Vilcanota depression is in that part of the southern Peruvian Highlands where the major physiographic alignments are broadly oriented in a northwesterly-southeasterly direction (Fig. 2). To the east of the valley is the Vilcanota Range which extends northwards from the massif area known as the Vilcanota Knot. Further north the Paucartambo River Valley separates the Vilcanota Range from the Pau-

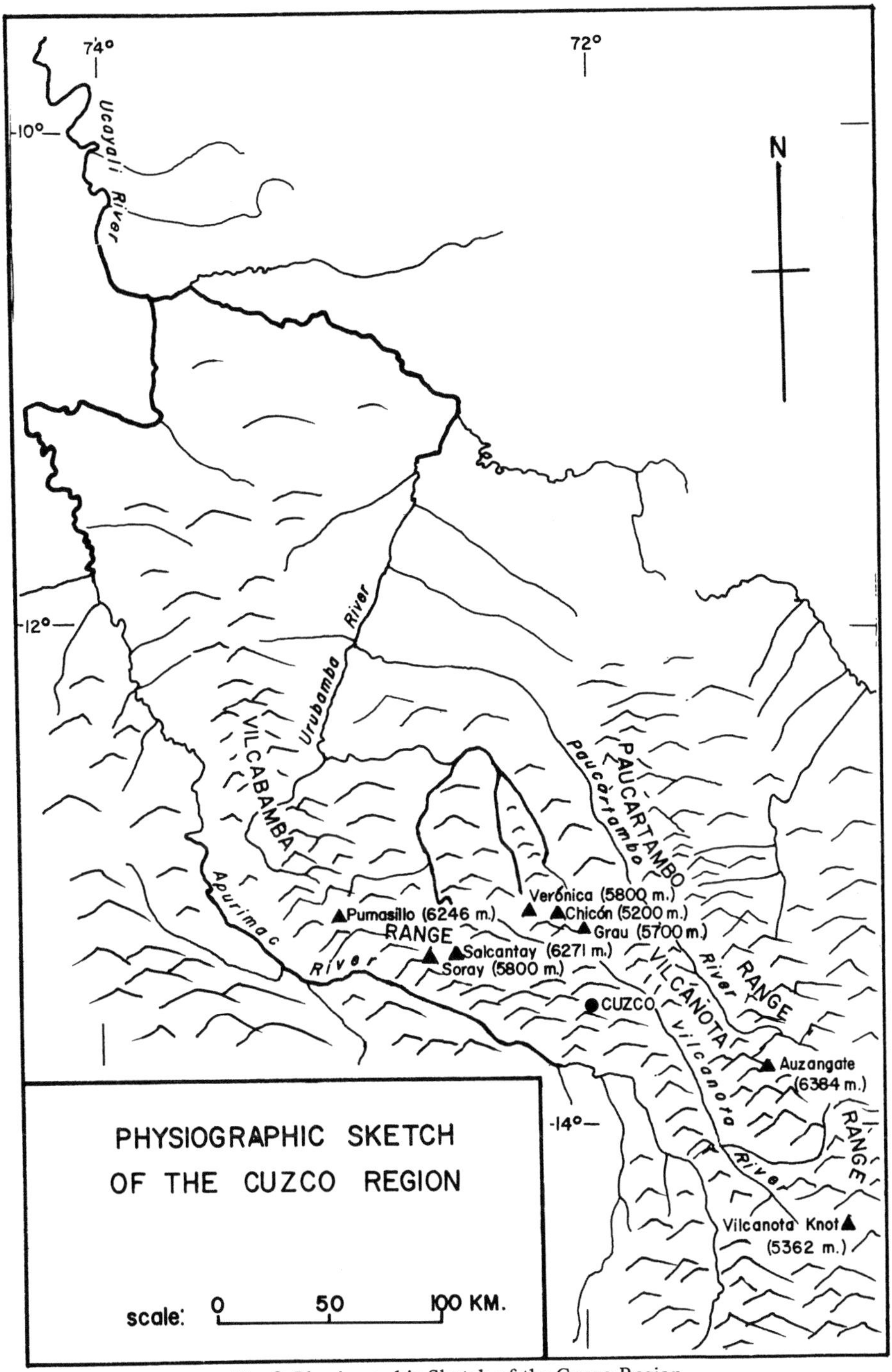

Fig. 2. Physiographic Sketch of the Cuzco Region.

cartambo Range. The southern Vilcanota Range has several beautiful snow-capped peaks: Ausangate (6,384 m), the highest mountain in southern Peru; Grau (6,700 m) and Chicón (5,200 m) both near Urubamba; and Verónica (5,800 m) north of Ollantaytambo. In the vicinity of Machu Picchu this range crosses the Vilcanota Valley and becomes known as the Vilcabamba Range which contains the impressive snowcaps Salcantay (6,271 m), Soray (5,800 m) and Pumasillo (6,246 m).

Another range extends northwards from the Vilcanota Knot and separates the Vilcanota and Apurímac depressions (GADE & ESCOBAR, 1972). It is a much lower range of mountains than the Vilcanota Range to the east of the valley and few peaks reach above 4,000 m. Northward it merges with the Vilcabamba Range.

HYDROGRAPHY

The Vilcanota Valley has been formed by the Vilcanota River which, fed by glacial meltwaters, several lakes and rainwater, flows in a general southeast to northwest direction (Fig. 3).[1] The river has its origin in the snows of the Vilcanota Knot at 5,362 m where the meltwaters unite in a tiny lake at 4,314 m[2]. These waters travel many thousands of kilometers to the Atlantic Ocean by way of the Vilcanota, (Lower) Urubamba, Ucayali and Amazon Rivers, increasing in volume from the innumerable affluents along the way.

The most important affluent in the study area is the Salcca River which has its source about 100 kilometers to the east. The Salcca, when it enters the main stream, deepens the Vilcanota by at least 2 meters, but does not noticeably widen it. The streams from the west side have much less flow than those on the east side because the mountains are not so high and do not have permanent snow on them until the Vilcabamba Range is reached. The Vilcanota River is about twice as deep in the rainy months of January and February as it is in July, August and September. In the low period, the water is clear but during the rainy season the river becomes chocolate-colored from the sediment, much of it valuable topsoil, washed into the river. Water from the river may be lost through evaporation but not by removal for irrigation; irrigation water always comes from streams flowing into the main stream, never from the Vilcanota itself.

[1] The two most common names for this river are Vilcanota and Urubamba, but Vilcanota is preferred because it is the one used by the Peruvians themselves. Vilcanota is a Spanish corruption of 'Wilkamayu' which is Quechua for 'river of the area where the *wilka* tree grows', a semitropical species found below 1,700 m in the valley. Technically, the Urubamba begins at the confluence of the Yanatile and the Vilcanota. Not until the twentieth century did the name of either Vilcanota or Urubamba become standardized throughout the area of the study. In his travels through the valley, MARCOY (1875:158) noted that the river changes names frequently from place to place taking the name from the important towns on the banks; between Quiquijana and the town of Urubamba, the name changed six times.

[2] In September, 1907, BOWMAN (1916:276) found the snowline here at 4,970 m.

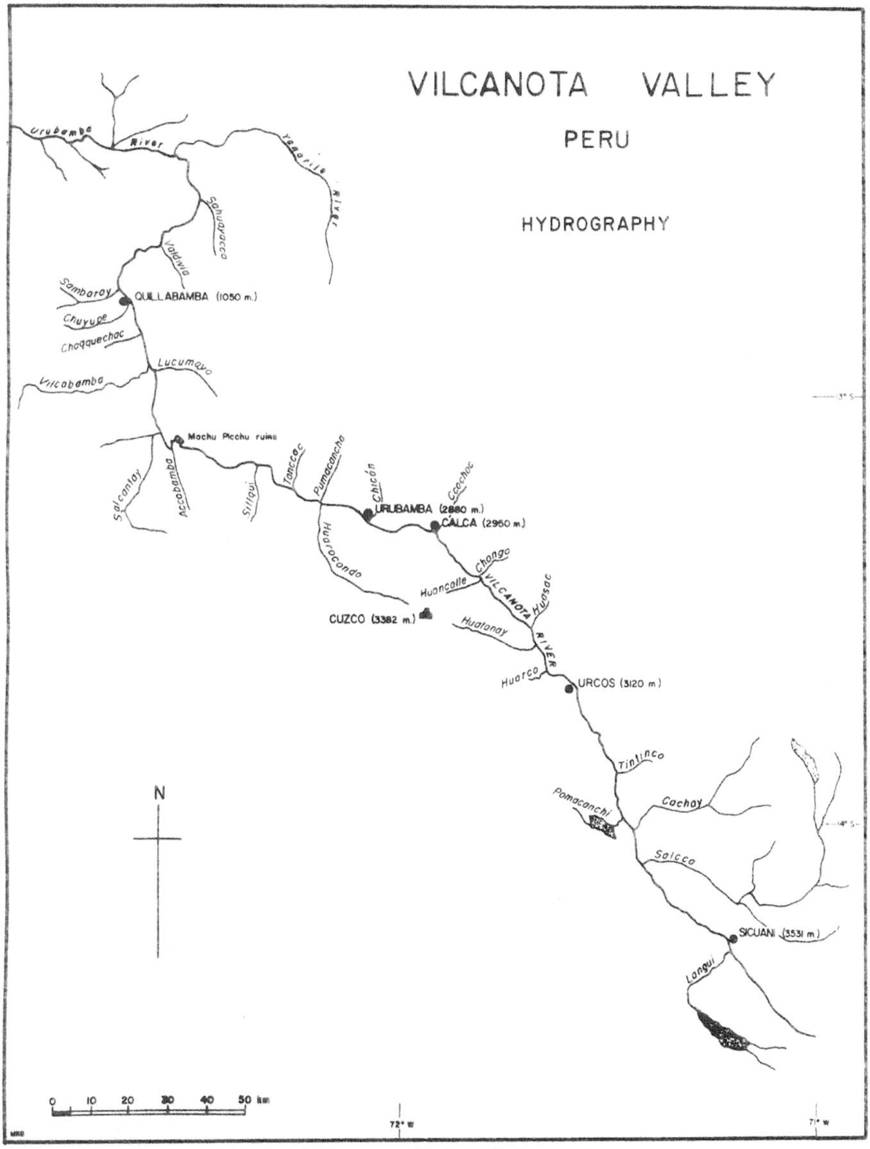

Fig. 3. Vilcanota Valley: Hydrography.

LANDFORM

The Vilcanota Valley has a complex geomorphology for not only have several agents of erosion been at work in forming the depression, but also the valley lies in a zone of contact between many different kinds of rock. For about 65 percent ot its length, the valley has a U-shaped profile with steep sides and a floor more flat than hilly. Whether glacial ice or running water was primarily responsible for this form is uncertain. With a few exceptions this form is maintained between 4,314 m at La Raya Pass (where the valley floor is so flat that it is marshy) and 2,400 m. The average width of the valley floor between these elevations is 3 kilometers. In places the river has cut down below the flat floor, incised as much as 15 meters, but it is more usual for the river to meander slowly on the surface. The river terraces, from 2 to 100 meters above the channel, may be virtually flat or quite rolling with as much as a 15 percent slope. In large measure this section of the valley has folded sedimentaries as its bedrock, but at an elevation of 3,060 m the river has cut through a basaltic mass. At those junctures where lateral valleys meet the Vilcanota depression, extensive flat areas highly adaptable to agriculture are found (Fig. 4).

Fig. 4. View of the Valley looking downstream from Ollantaytambo, around, 2.780 m. The Valley bottom here is unusually wide.

Approximately 10 percent of the valley distance, between 2,400 m and 1,500 m, is a narrow canyon with an extremely steep gradient where the river has cut through the Vilcanota-Vilcabamba Range (Fig. 5). At Torontoy (2,400 m), the valley narrows as the Vilcanota River flows between massive granite slabs rising perpendicularly more than 900 m for most of the distance of 30 kilometers. At an elevation of 1,500 m the river emerges from the canyon, having maintained for this distance a gradient of 30 meters per kilometer (Fig. 6).

From Hacienda Ccolpani at 1,500 m to the junction with the Yanatile at 730 m – about 24 percent of the valley distance – a fairly broad valley[3] has been formed by the work of running water. The valley sides commonly have a 20-35 percent slope, and from 60 to 150 meters above the river there are many river terraces (*mezadas*) which are favored places for human settlement (Fig. 7), such as the one on which is situated the town of Quillabamba.

Alluvial fans are found in all sections of the valley ranging in size from several square meters to several square kilometers in extent. The large fans have been formed by stream aggradation in the mouths of the valleys that enter the principal valley. Some of these fans have a steep gradient, are covered with large boulders and cut up by distributaries and are usually left uncultivated. Many fans, however, are gently sloping, of fine rather than coarse materials and are used for agriculture and settlement sites, e.g., at Santa Teresa and Urubamba. Most of the man-made agriculture terraces in the valley were constructed on alluvial fans. It is debatable to what degree these elaborate, man-made constructions were a conservation measure to stabilize the talus slope, and to what extent they were a means of adding more crop land to feed an expanding population.

SOILS

Soils in the valley are generally residual on the slopes and alluvial on the valley floor. Beyond this, few valid generalizations can be made regarding the nature of the soils. Residual soils of the upper slopes are derived from either sedimentary,

[3] The local inhabitants make an essential distinction between 'quebrada' and valle' using a definition of these terms beyond a strict landform designation implied in general Spanish usage. 'Quebrada' (gorge) is applied to lands above 2,400 m in the depression and has a temperate climate, whereas 'valle' (valley) refers to the more open trough below 1,500 m with its tropical environment. The canyon-like area between 2,400 m and 1,500 m is often called 'medio valle' (literally, half valley). All parts of the depression are, of course, 'valley' in the geomorphological sense. In this study, 'lower valley,' unless qualified, refers to that part of the study area which is below 1,500 m while 'upper valley,' unless qualified, refers to that part above 2,400 m. Likewise the term 'tropical' will refer to the environment below 1,500 m, even though in a mathematical sense, all of the valley would be tropical because it lies between the Tropics of Cancer and Capricorn.

Fig. 5. The canyon of the Vilcanota at an elevation (at the bridge) of 1,910m.

igneous or metamorphic rocks. Sandstones are common above 2,600 m and slates below 1,500 m elevation. Soils of granitic origin are found below 2,600 m to about 1,500 m and are blackish in color. A tendency towards iron oxide concentration begins below 1,000 m elevation, but classical tropical lateritic soils do not appear until farther down the Urubamba at 500 m above sea level.

Residual soils are generally thin with little profile development because of constant sheet erosion. Transported soils in the valley are found on the floodplain, river terraces and on alluvial fans. The lower terraces generally have clayey soils of heavier texture and higher calcium content than the more loamy soils of the upper terraces. At the level of the stream channel itself, the soils are sandy with much gravel and cobble; these areas are frequently left uncultivated. The soil on the valley floor in much of the depression is still highly productive after centuries of constant use.

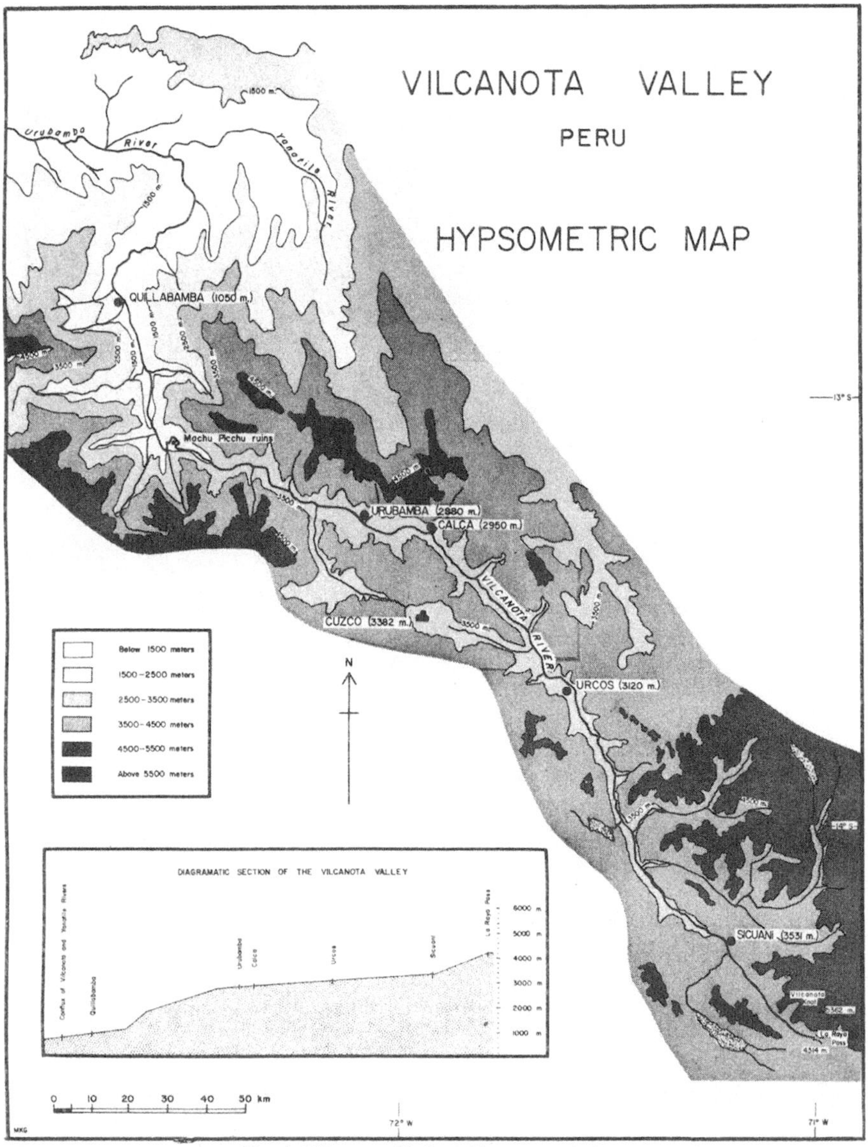

Fig. 6. Vilcanota Valley: Hypsometric Map.

Fig. 7. A high-level river terrace (*mezada*) cut through by a stream, and with an alluvial fan at its base (1,500 m).

CLIMATE

Only fragmentary and unreliable climatic data are available for the Vilcanota Valley and it is impossible to deal with climatic differences on an accurate quantitative basis[4] The only weather records for the whole department regarded to be fairly reliable are those for the city of Cuzco, 30 kilometers west of the valley at an elevation of 3,382 m. Based on an 8-year average, the average annual precipitation is 800 mm., the January temperature average is 11.38° C. and the June temperature average is 8.99° C.[5] The most apparent climatological generalizations for the valley are the distinct concentration of precipitation during the high sun period, and the increase in average monthly temperatures with a decrease in elevation above sea level.

[4] Prior to 1960, only one meteorological station had been operated in the valley, that of the Salesian school near Yucay, but the reliability of the data collected at this station was regarded as very low by the director of the school himself. In 1963 the Ministry of Agriculture established a Center of Agrometeorology and Hydrology to set up weather stations in some of the more important towns in the department; unfortunately, even the very short-term and fragmentary data from these stations is considered to be unreliable because of the inexperienced observors who collect and record it.
[5] Data from the meteorological office at the Cuzco airport.

Throughout the valley and the whole southern Sierra the rainy period comes from October to April; although the highest temperatures of the year occur during this season, it is nevertheless called 'winter' (*invierno*). The actual annual precipitation varies in amount from 400 to 1,800 millimeters, but most of the valley may be classified as subhumid if the term is defined as rainfall effectiveness. Only the *ceja de la montaña* (2,400 to 1,500 m) may clearly be considered to have enough (and indeed too much) rainfall for effective crop growth without irrigation. The smallest amount of rainfall is found in the southern part of the valley in the Province of Canchis which is also the zone of most frequent droughts. From field observations and by interpolation of the fragmentary data that is available, the following climatic regions may be, *faute de mieux*, delineated (Fig. 8):

1.MICROTHERMAL ZONE. This zone extends from 4,314 m to 3,900 m. In June and July the average monthly temperatures here are below 5° C. and no month averages above 10° C. Ice forms at night during half of the year but melts during average above 10° C. Frosts begin to occur in May, and in June temperatures may tion in most years is below 400 millimeters.

2. COOLER MESOTHERMAL ZONE. This zone extends from 3,900 m to 3,000 m and is the most important climatic belt in area. From one to five months average above 10° C. Frosts begin to occur in May, and in June temperatures may drop as low as -7° C. at night. Annual precipitation is around 500 millimeters.

3. MIDDLE MESOTHERMAL ZONE. This zone extends from 3,000 m to 2,400 m and differs from the previous zone mainly in its slightly warmer temperatures. However, frosts frequently occur in the low-sun period in the upper part of the zone. Rainfall amounts to 550-600 millimeters.

4. WARMER MESOTHERMAL ZONE. The climate in this zone could best be described as rainy subtropical. Only very rarely is the temperature below 0° C. in this zone, but equally rare are temperatures above 15° C. The Vilcabamba Range effectively acts as a barrier to hot air from farther down the valley. This zone is characterized by an almost constant drizzle for about half of the year. From November to April, at least some rain falls 25 days out of every month. In each of the dry months, May through October, six days are rainy. In spite of the frequency of precipitation, the total amounts (*c*.2000 millimeters) are not actually much greater than those received farther down the valley. Since the temperatures are lower here than in the following zone, the rainfall has a higher effectiveness for the growth of vegetation.

5. MACROTHERMAL ZONE. This zone extends from 1,500 m above sea level to 730 m (lower limit of study area) and beyond. The monthly average is in all cases above 21° C., but the nights are remarkably cool. In some years the precipitation may be as little as 1,000 millimeters but more typically the amount is between 1,500 and 1,800 millimeters. About 60 percent of the rain comes between December and March.

Certain weather elements not apparent from temperature and rainfall records are important in this valley in so far as they have an effect on the growing of plants. The short days, combined with the relatively low average temperatures above 2,500 m, make for slow crop growth and maturation. Whereas in the mid-

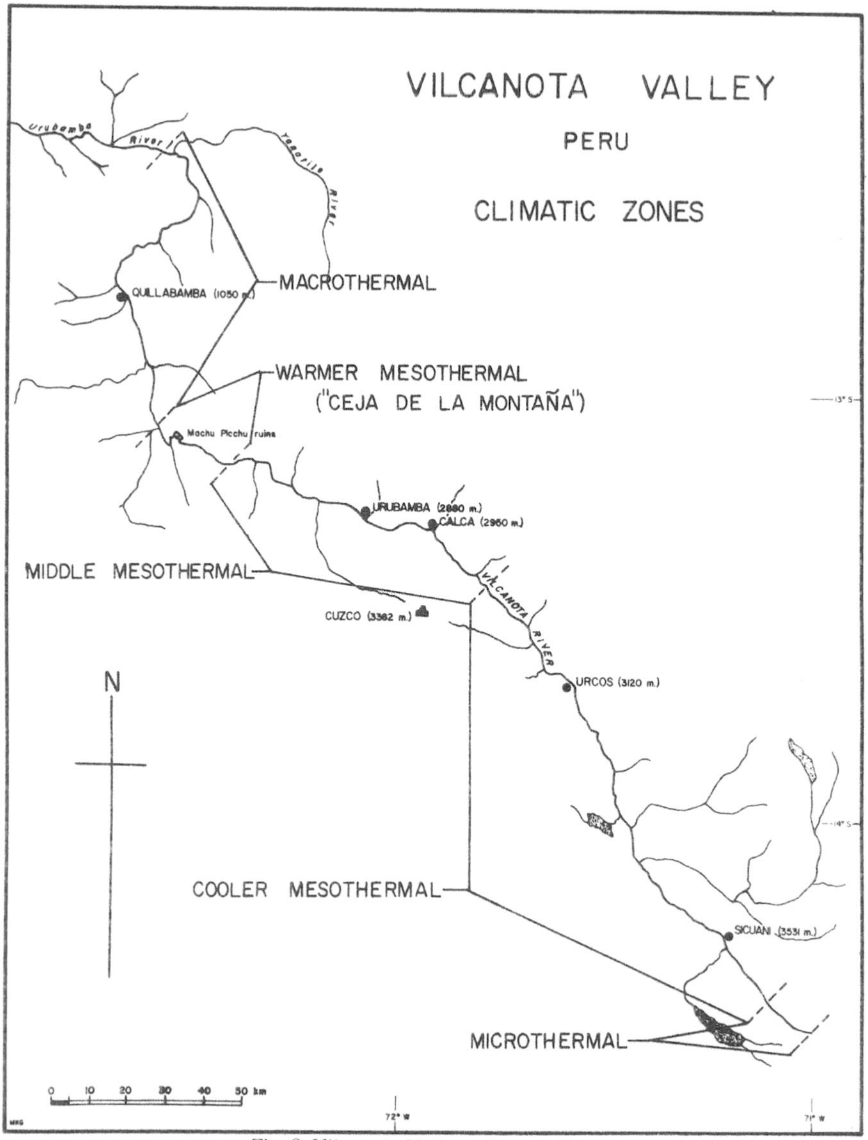

Fig. 8. Vilcanota Valley: Climatic Zones.

latitude summer maize may complete its growing cycle in $3\frac{1}{2}$-4 months, in the southern Peruvian Highlands the growing season is from 7-8 months above 2,500 m.

Hail, snow and dew are common in the valley. Hail is frequent above 3,400 m and liable to occur as low as 2,500 m. A possible hailstorm ('granizada') causes great anxiety among the peasants for it can completely destroy crops in a matter of minutes. Snow occasionally falls and can damage or destroy growing crops. A notable snowfall in the colonial period struck the area on May 1, 1699; it lasted 3 days and destroyed crops and animals (MONTESINOS 1906:198). Dew can have either a positive or negative effect on agriculture. Ripe grain which shatters easily is often harvested in the early morning when the dew on the stalk prevents the grain from falling to the ground. Dew, however, can also act as a magnifying glass, causing incoming solar rays to seriously burn the plant tissue; to prevent this, a rope is stretched across the field and moved along to shake the dew off the stalks and leaves.

VEGETATION

The vegetation of the Vilcanota Valley becomes increasingly complex, both in species and physiognomy, as one moves from the microthermal zone in the south to the warm 'valle' in the north. Above 3,910 m the *puna* landscape prevails consisting mainly of coarse bunch grass growing in tussocks. *Stipa*, the most important grass genus here, covers the valley floor and some of the slopes. Bromeliaceous plants such as those of the genus *Puya* are also common on many slopes at this elevation. Below 3,850 m the valley floor has been cleared for agriculture and little wild vegetation remains. The upper slopes on the valley sides are covered mostly with native grasses, while the lower slopes are covered with herbs and shrubs. Uncultivated native trees are conspicuously absent and now grow only in some of the less accessible lateral valleys. Cacti, eucalyptus, and the kishwar tree (*Buddleia*) are the most common non-crop plants below 3,750 m. In poorly drained areas, rushes (*Juncus* spp.) and grasses (*Paspalum* spp. and *Bromus* spp.) dominate. Between 3,300 m and 2,500 m woody shrubs are common on slopes, along river banks and on roadsides; among the most ubiquitous are *Baccharis* spp., *Spartium junceum, Schinus molle, Eupatorium* spp. and *Cassia* spp. The vegetational landscape of the upper valley reflects the highly seasonal rainfall. Between January and April verdure predominates, but beginning in May, yellow begins to take over. In June and July the drought and frost turn almost all vegetation brown and this color prevails until late October.

Around 2,500 m the plant cover becomes noticeably denser and some tropical elements from the lower valley enter the flora. At 2,450 m, near the river, the vegetation almost suddenly becomes thicker and higher with the appearance of shrubs taller than 2.5 meters. At. 2,300 m the ground begins to be covered with turf, trees dramatically appear and the tropical-looking, moisture-loving species are apparent. Cactus drops out. At this elevation, some slopes are forested while others are still herbaceous in character; this pattern is not explained by slope exposure. Beginning at 2,340 m, the *pisonay* tree (*Erythrina falcata*) forms a solid riverbank association down to approximately 1,800 m. At 2,200 m one is in the *ceja de la montaña*, perhaps the densest vegetation mantle known anywhere. Many epiphytes, particularly orchids and bromeliaceous plants, as well as tree ferns and mosses grow on, in or under the evergreen broadleaf trees. Bamboo (*Chusquea*) is common on all disturbed hillsides and is the major hindrance to human movement in this area.

At 1,500 m the valley widens and the atmosphere is drier; cactus is again a component on some lower slopes and a sparse savanna (*pajonal*) covers the dry hillsides. On hilltops that have not been cleared a denser tree cover is found because they are exposed to the moisture in the low clouds. The forest extends to lower elevations on shady slopes than on those which face the sun most of the day; many of these forests are in a secondary-growth stage (*purma*). The vegetation in the valley bottom is almost xerophytic in character until around Echarati (965 m) where the area near the river has more tree growth. However, one does not enter the rainforest formation until near the juncture of the Yanatile and Vilcanota. Near the river at this juncture the canopy forest has been cut and what exists is largely second growth.

The present character of natural vegetation in much of the Vilcanota Valley is not primarily an ecological response to the forces of nature. Temperature, rainfall or soil conditions cannot basically explain the absence of tree vegetation in most of the middle and upper sections (2,550-3,750 m) of the valley, although these factors may help explain the slow natural regeneration. Nowhere in this aforementioned area is the precipitation less than 400 millimeters, an amount sufficient for tree growth unless the soils are unusually porous. Nevertheless A. WEBERBAUER in 1911 (1945) – the first scientifically trained botanist to study the valley – saw the vegetation to be a reflection of non-human ecological factors. Later, O.F. COOK (1916a:292) refuted this view and interpreted the present pattern as clearly being man-induced. Among his evidence was the natural reforestation on now uncultivated Inca terraces near Ollantaytambo. An area densely populated for at least a millenium must have been gradually stripped of its forest cover starting with the valley floor and progressing up the slopes. The motivations were several: clearing of land for agriculture, collecting firewood, and the extension of pasture land which increased with the introduction of Old World livestock after 1532. In the sixteenth century wood was already scarce and a penalty of 100 lashes was administrated to any Indian who cut down the main trunk of a tree (LEVILLIER, 1925:VIII:371). The shortage of wood prompted the Spanish authorities to require the planting of willow, poplar and capulí trees along irrigation ditches and on stream banks (*Ibid.*). In the latter part of the nineteenth century eucalyptus was introduced and is now by far the most common tree species in the valley. Planting trees in the Peruvian Sierra to insure an accessible supply of wood is apparently at variance with the pattern elsewhere in Highland Latin America of simply seeking wood needs farther afield from the wild vegetation (DICKINSON, 1969:294).

Human-induced physiognomic influences are apparent within the valley. Above 2,400 m, much of the present natural vegetation is pioneer in character and only those shrubby or herbaceous plants which have adjusted to disturbed situations survived as important components of the landscape (Fig. 9). The upper Vilcanota Valley has been a major source of firewood for the city of Cuzco (HARDY,

Fig. 9. Wild vegetation, mainly bromeliaceous plants above and molle trees on the lower slope, near Andahuaylillas (3,100 m).

1914:503), and it is not difficult to imagine how this affected the woody plants that grew natively on the slopes.

In the zone between 2,400 m and 1,500 m, about 60 percent of the forest cover has remained intact because of inaccessibility of the steep slopes, low human population density and high regeneration rate. In the 1870's a sawmill was established in the ceja at Torontoy, the beginning of selective tree cutting in this zone (Memoria Urubamba, 1874:385). From 1,500 m down to 600 m, most of the forest has now been cut over, but because the regeneration rate is high, woodland predominates. As late as 1840, the area below Echarati (960 m) is said to have still been in virgin forest.

Evolution of Human Occupance in the Valley

The antiquity of man in the Vilcanota Valley is not known and even the approximate time that sedentary living began is highly conjectural. Based on the interpolation of data from other parts of the Central Andean region, man may have first appeared in the valley as a hunter and gatherer and this way of life may have formed the basis of his subsistence for about eight millennia. Certainly by 1000 B.C. (BENNETT, 1946:142), he had become a sedentary agriculturalist and by A.D. 500 the valley was probably well occupied by several different cultural groups who practiced irrigation, kept llamas, and lived in semidispersed settlements. It was not until *c* A.D. 1100 that the Incas, one of the tribes to the west of the Vilcanota Valley, started to expand out from their original core in the Cuzco Basin. The Incas established hegemony over adjacent groups and probably by A.D. 1200 all of the highland groups in the Vilcanota Valley had been conquered by the Inca, politically absorbed into their expanding empire and culturally absorbed by adapting their Quechua language and sun-cult religion. Some groups seem to have been more tenacious in maintaining their cultural identity than others, e.g., the Canchis people in the present-day Province of Canchis. By A.D. 1475 one may assume that all of the valley above 1,500 m elevation was effectively incorporated into the Inca state and permanently settled by Quechua-speaking peoples (ROWE, 1946:205). After A.D. 1400 at least part of the area below 1,500 m came under the control of the Incas as a coca-growing region (ROWE 1946:208).

A large literature on the material and non-material configuration of Inca culture is available.[6] The Inca civilization was solidly based on agriculture, with the vast majority of the subjects of the empire engaged in growing crops. The Inca nobility exacted tribute in the form of agricultural products such as quinoa, maize, dehydrated potato and coca, which represented specialities of different environments. The Vilcanota Valley played an outstanding role in the development and elaboration of Inca civilization. It was a major food supply area, a bulwark against real or imagined invasions from outside the Empire, a playground for those escaping the cold of the capital, a center for ceremonial activities and an important transportation route linking the capital to the Titicaca area and points south (GADE, 1968). The number and variety of architectural ruins in the valley testify to the importance of Inca settlement in the depression above 1,500 m (Fig. 10). The ruins, together with the natural capabilities of the land, make it not unreasonable to estimate that at least 150,000 people inhabited the valley in A.D. 1500, but no more than 2,000 of these lived below 1,500 m at any one time.

Ruins are found throughout the Canchis section of the Valley, but the most impressive ones are near Racche where the Incas constructed the Wirakocha temple. Walls built of stone ashlars in the lower part and of adobe in the upper are all that remain of the temple itself, but many enclosures constructed of the native lava rock surround the temple. At La Raya Pass (4,314 m) the *tambo* (wayfarers post) and fortifications that marked the division between the Canchis people and the Kolla people of the Altiplano (SQUIER 1877:400) have now disappeared.

North of Racche and the Wirakocha temple, the outstanding site is Pikillakta which, with its broad esplanade in the middle of a village complex, may have served as a marketing center for the Incas. Most of the other ruins between Wirakocha and Pikillakta are on hills rather than in the valley bottom itself. North of Pikillakta are the ruins of Pisac. Here on a high ridge above the valley is combined a fortress and a temple with an elaborate terrace system extending down the slopes

6 The outstanding survey of Inca culture in any language is undoubtedly that of JOHN H. ROWE in the Handbook of South Americans. More recently, JOHN HEMMING (1970) has superbly treated the history of the late Inca period and the Spanish Conquest.

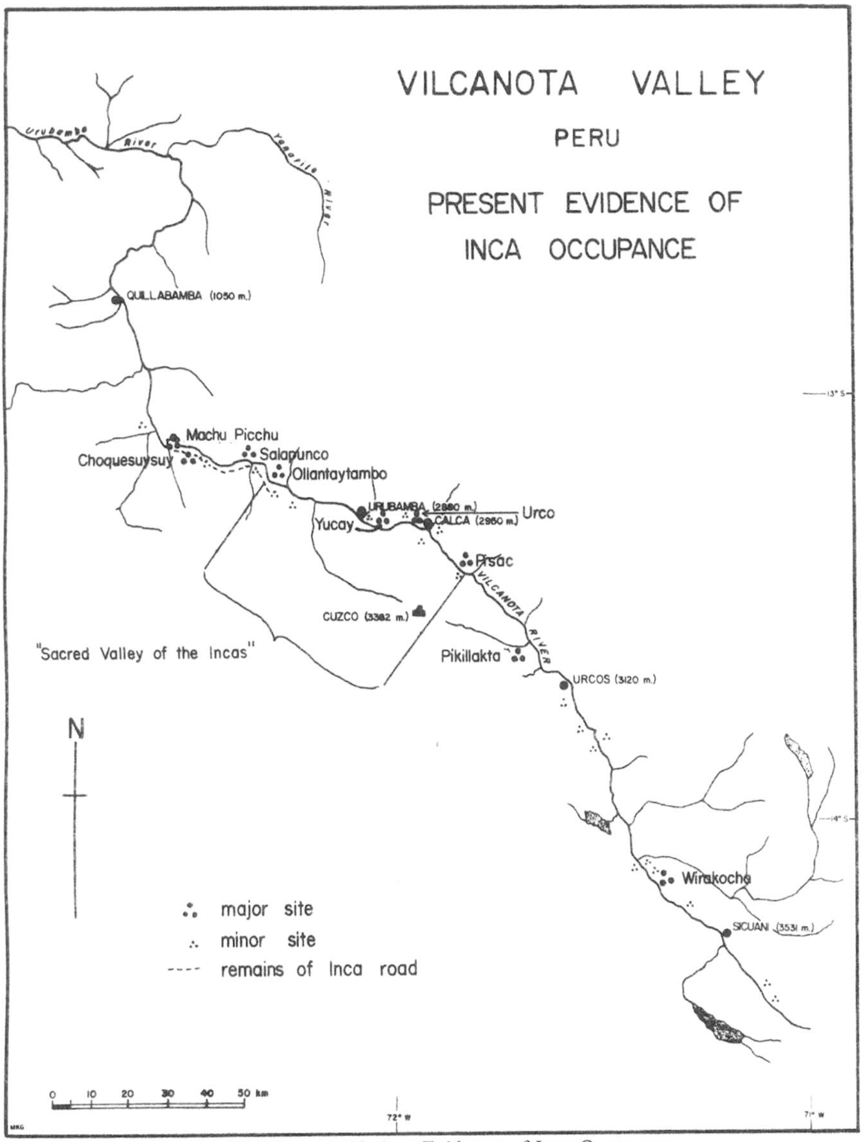

Fig. 10. Vilcanota Valley: Evidence of Inca Occupance.

toward the Vilcanota River. Solid stone walls, constructed with a backward inclination and staggered in the form of steps, held the soil to grow food necessary to sustain the warriors and priests in the event of a siege. Near the present village of Pisac the Incas straightened the channel of the Vilcanota by constructing rock

Fig. 11. Ruins of the Inca city of Machu Picchu, 600 meters above the Vilcanota River.

walls along the river banks, thus preventing stream meanders and the resultant degradation of high quality farming land.

Between Pisac and Ollantaytambo, the valley is commonly given the name 'Sacred Valley in the Incas' because of its rich associations of agriculture and settlement during the Empire period. At Urco near Calca is some comparatively little-known but impressive evidence of Inca occupation; at Yucay there are well-constructed terraces, cemeteries, and remains of important structures. At Ollantaytambo a fortress occupied a hill about 1 kilometer from the river with the dwellings clustered below. Farther down the valley were several well-made structures with agricultural terraces (now abandoned) such as Salapunco (2,500 m) and Choquesuysuy (2,200 m). Just before the Vilcanota River makes a horseshoe bend are the ruins of the spectacular city of Machu Picchu situated *c.* 600 m above the level of the river. This virtually impregnable fortress settlement had an important religious function, at least 100 dwellings and many terraces which extend down some extremely steep slopes (Fig. 11). These are of finely constructed stone walls up to 2 meters high, about 80-100 meters long and averaging 8 meters deep.

The Vilcanota Valley from its head to about Ollantaytambo was an important natural passageway for the movement of men, goods and ideas in the Inca period. *Tambos* were established to aid travelers and at least some of the villages in the valley may have begun as rest stops. North of Ollantaytambo, an Inca trail followed the Vilcanota gorge, partially in it and partially above it, to Machu Picchu

Fig. 12. The last evidence of Inca occupation toward the lower valley is this small terrace complex at 1,830 m.

and below this city to about 1,800 m; below this elevation there is no evidence of such a trail.

Pre-Conquest occupance of the Vilcanota Valley below 1,500 m remains an enigma. Known to the Incas as the *Yunca* (or *Yunga*), this territory conspicuously lacks Inca monuments and thus contrasts significantly with areas above this elevation where architectural remains abound.[7] It is very possible that the then unhealthy, hot environment precluded permanent settlement as the Incas perceived it in this and similar valleys. There can be little doubt that this part of the valley was valued by the Incas not only for coca but also for growing certain fruits such as *cherimoya* and certain root crops such as *achira, llacón* and possibly manioc. CIEZA DE LEÓN (1883:49-50) noted that the Incas sent workteams (*mitamaes*)

[7] I. BOWMAN (1916:77) claimed to have found Inca stone terraces for growing coca and patios to dry the leaf at Echarati (960 m) and Sahuayaco (840 m), but personal field investigation failed to support this observation. In the valley below 1,500 m are many crudely constructed agricultural terraces which have nothing to identify them as being of Inca origin. However, Inca settlement ruins exist high above the valley in a cooler, more temperate environment which may have been beyond the reach of diseases such as those indicated above. COOK (1925:108) found a line of 60 Inca storehouses at Pinyasnioc near Panticalla (Málaga) Pass between the Vilcanota and Lucumayo (Amaybamba) Valleys at an elevation of 3,600 m. These storehouses may have been placed here to supply the lower valley with maize and potatoes from the upper valley. BINGHAM (1913:530) found ruins of another group of storehouses at an elevation of around 3,000 m above the main valley from Santa Teresa.

in shifts to the hot valleys to sow maize and grow coca and fruit which could not be grown in higher, colder environments. The Inca fear of diseases, such as leishmaniasis and yellow fever, may have prompted a continual back-and-forth migration of peoples which explains the absence of monumental-type architecture (Fig. 12). It is unlikely that the Incas obtained coca by trade from the poorly organized forest Indians on a regular basis.

The forest Indians who have lived on the northern fringes of the study area within historic times were the Machiguenga, a subtribe of the Arawak-speaking Campa group. Today they have been pushed deeper into the tropical forest of the lower Urubamba Valley and no longer have any settlements on the Vilcanota River upstream from its juncture with the Yanatile River (GADE, 1972b). About 1,500 of these Indians remain, very few in their aboriginal state, and about 70 percent of them have been Christianized by Spanish Dominican priests who maintain their regional mission headquarters in Quillabamba. Traditionally the Machiguenga have practiced a seasonal cycle of fishing in the dry season along the river, and slash and burn agriculture in the hills above the river during the wet season. Although trade was established between these lowlanders and the highland people perhaps as early as the pre-Inca period, relatively few cultural traits have been exchanged despite the proximity of the two groups.

POST-INCA PERIOD

The Spanish Conquest caused a geographical reorientation in the valley beginning soon after Pizarro's triumphal entry to Cuzco in 1553. *Encomiendas*, the right to use the labor of Indians living within a certain area, were granted to men who participated in the Conquest; these grants included the districts around the present-day towns of Maranganí, Tinta, San Pablo, San Pedro, Combapata and Checcacupe, as well as the huge 'Marquesado de Oropesa' which extended from about Pisac to Ollantaytambo. (Fig. 13). Tribute was exacted from the Indians living within these *encomiendas* and extra-legal work services were demanded. Tribute was nothing new to the Indians for they gave it in Inca times as well.

In 1545 silver was discovered in Potosí in present-day Bolivia, and by the seventeenth century the more than 200,000 Spanish administrators and Indian laborers of this *puna* mining center depended on outside areas to provide all their food. The Vilcanota Valley helped supply Potosí with some provisions such as dried fruit and coca (ACOSTA, 1962:181) which were carried there by llamas and mules.

Beginning in 1570, a massive resettlement of the indigenous population drastically changed the settlement pattern of much of the Vilcanota Valley and the whole Central Andes. From the dispersed and semi-dispersed settlement of Inca times, the Indians were moved into clustered settlements (*reducciones*) where they could be more easily managed by the civil and religious authorities. Most towns in the valley today originated as *reducciones*, and were established on the valley floor in the midst of valuable agricultural land such as at Pisac (Fig. 14). In

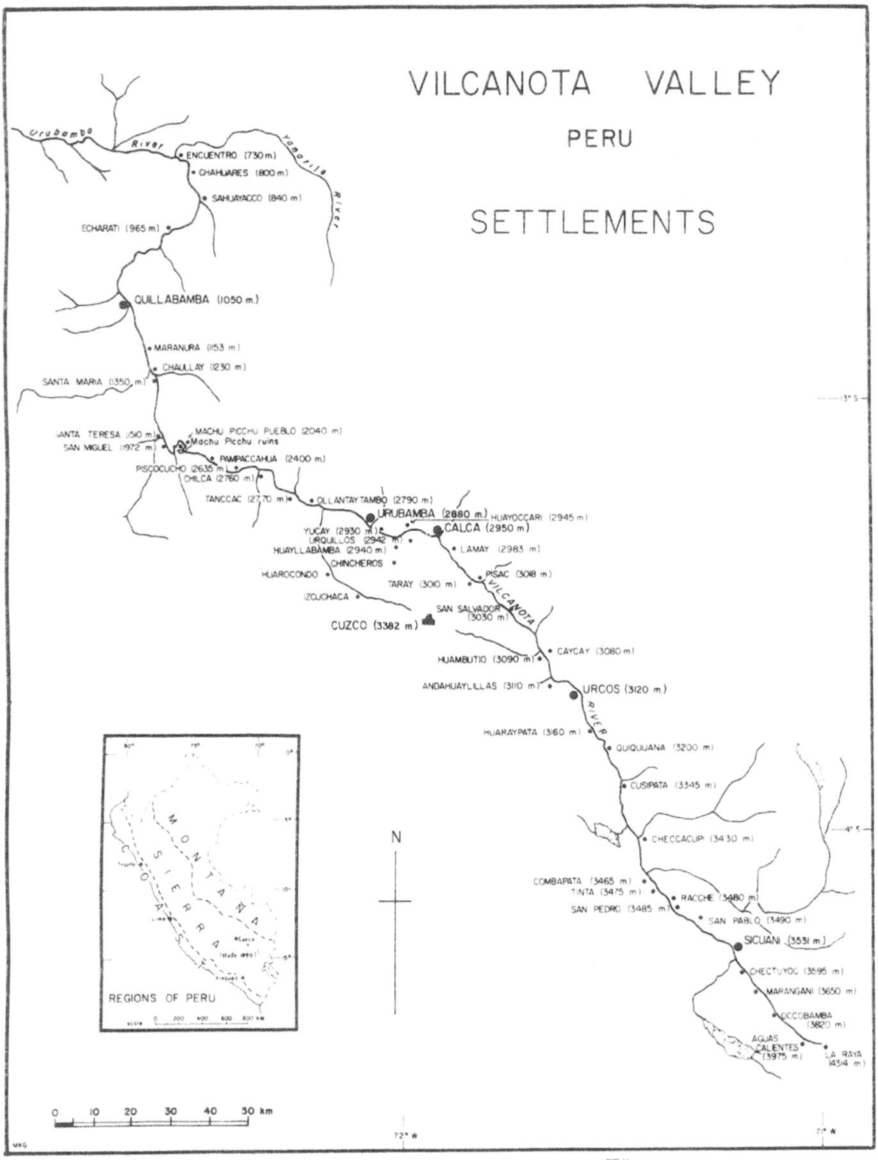

Fig. 13. Vilcanota Valley: Settlement; (Inset map shows regions of Peru and location of study area).

the sixteenth century Viceroy Toledo promulgated a multitude of edicts affecting these towns; one stated that, except for small gardens near the dwellings, no fields should be on the immediate periphery of the towns because of health changes

Fig. 14. The post-Conquest settlement of Pisac, set in the valley bottom, below the Inca ruins on the hill far above (3,018 m).

caused by 'increasing the humidity' (LEVILLIER, 1925a:165). Indians from other areas were brought in to settle in the valley and one may assume that new varieties of useful plants and agricultural customs were brought also to the valley during this transfer.

Interpolation from early colonial population records (VÁZQUEZ DE ESPINOSA, 1942:709-711) suggests that the population of the valley in 1600 may have been 90,000 or about three-fifths of the estimated total for 1500. The introduction of Old World diseases and the transfer of native laborers to the Potosí silver mines caused this population decrease which was most noticeable in the southern part of the valley. Canchis and Quispicanchis (from La Raya Pass to Urcos) were the areas in the valley – and among the 16 provinces in the *Audiencia* – from which the Indians were forceably recruited for the mines, taking with them their whole families.[8] By 1700 (KUBLER, 1946:338) the population of the valley may have further decreased to around 50,000. In 1800 the figure may still have been 50,000 for the valley, although this estimate is based on a very fragmentary source (GONZÁLES, 1833).

At the end of the sixteenth century, *corregimientos* took over from the *encomiendas*. These were properties owned by the King of Spain and were designed to

[8] Lurucachi, a town in southern Canchis, which was often mentioned in the early records has essentially disappeared, although the place name is still used.

protect the rights of the Indians by assuring them the use of the land, although technically they no longer owned it. In the seventeenth century, however, land began to drift into the hands of individuals or organizations and *haciendas* or manor estates were established on which the Indians were tenants ('yanaconas'). The Church acquired numerous large land holdings in order to support its religious activities. Particularly strong were the religious orders – Jesuits, Dominicans, Franciscans, Recollets, Mercedarians and Augustinians – as well as several convents of nuns. In the Vilcanota Valley the Society of Jesus acquired more land than any other group and had extensive holdings both in the upper and lower parts of the depression. Individuals also acquired much land, part of it by unjust exploitation of the Indian.

Indian communities, particularly in the present-day Provinces of Urubamba and Calca, were left with the poorer agricultural land. Those indigenous communities that remained were forced to pay tribute to the civil authorities usually in products that the Indians cultivated on their own lands, but other tribute items were also accepted such as firewood or rope made from agave fibers. If an Indian community could not meet tribute obligations, its lands might be offered for sale (KUBLER, 1950:342). Those Indians who worked as peons on estates were not expected to pay tribute. The needs of administering tribute obliged the Spaniards in 1784 to restructure the political and economic division of the Viceroyalty of Peru, dividing it up into seven large units known as *intendencias,* one of which was Cuzco.

The Intendencia of Cuzco was broken up into 11 *partidos* (corresponding more or less to a county) which covered what are today the Departments of Cuzco and Apurímac; they were Tinta, Calca and Lares, Aimaraes, Cotabambas, Chumbivilcas, Paruro, Paucartambo, Quispicanchis, Urubamba, Abancay and (the valley of) Cuzco. Urubamba included all of the lower Vilcanota Valley and in the upper valley extended slightly past Huayoccari; Calca and Lares reached from there to Huambutío; Quispicanchis extended to around Combapata; and Tinta reached from about Combapata to La Raya Pass. Today the lower valley (below 1,500 m) is in the Province of La Convención which was created out of the Province of Urubamba in the 1930's. Calca (now without Lares) Province and Quispicanchis Province have maintained their positions. A tiny portion of the upper valley is in the Province of Paucartambo. The old Tinta unit had its name changed to Canchis, but its geographical position is essentially the same (Fig. 15).

The road system in the valley during the colonial period did not differ significantly from that of Inca times. The upper Vilcanota Valley road took on increased importance as a trade link from Cuzco to Potosí and probably was widened from the narrow trail of the Incas. The *tambos* were maintained to aid travelers and were often strategically situated as at La Raya Pass and at Lurucachi, the upper boundary of the 'temperate' zone of the valley (CONCOLORCORVO, 1959:357). Around each tambo, crops were planted to supply transients with food and forage for their horses and mules.

The Spaniards drastically altered the land tenure and settlement patterns of the temperate part of the valley from what they had been in the Inca period, and these

Fig. 15. Provinces of the Department of Cuzco.

changes involved a radical reorientation of peoples' lives. On the other hand, in the subtropical and tropical sections of the valley where few sedentary Indians lived, the shift to the colonial way of living was less dramatic. The ceja de la montaña, between 2,400 m and 1,500 m, did not interest the Spaniards because it was inaccessible, constantly rainy, and lacked desirable resources. Spaniards travelling to the lower valley did not follow the Vilcanota canyon but instead used the Málaga Pass and Amaybamba (Lucumayo) Valley. Brush had overgrown the Inca trail that led down the valley through the gorge and even if the Spaniards were aware of its existence, they did not use it.

Colonization of the lower section of the main valley began soon after the Spanish subjugation of the area around 1560. The first estate established in the valley proper was Hacienda Ccolpani Grande in 1560 which was soon followed by Hacienda Huadquiña in 1576 (MENÉNDEZ RUA, 1948:30). Not until 1650 did more of the Vilcanota Valley tropical zone open up; in that year, the Jesuits established Haciendas Santa Ana and Pintobamba Grande in the vicinity of present-day Quillabamba. In 1720, Hacienda Echarati, another Jesuit property, was started and in the course of the next hundred years the remainder of the lower valley was settled. As early as 1610 sugar cane mills (*ingenios)* and Negroes were mentioned (MAURTUA, 1906:I:239) indicating that Negro slaves must have been brought from the coast to the lower valley around the turn of the sixteenth century to work in the sugar cane fields. Hacienda Santa Ana was the center of agriculture and settlement in the lower valley for 200 years after 1650. This enterprise undoubtedly functioned as an important crop introduction center for the Jesuits who brought many seeds and plants to the lower valley from their estate near Abancay. Sugar cane cultivation became highly profitable and this estate became famous all over Peru for its sugar. In 1967 the Jesuits were expelled from the valley (and all of Latin America) and their estates were subsequently taken over by private individuals.

POST-COLONIAL PERIOD

During and shortly after the period in which Peru got her independence from Spain in 1824, the population declined because of the armed struggle, the concomitant social disorganization and from disease epidemics. The commercial agricultural structure which the Spaniards left in good working order broke down completely, and many of the estates throughout the valley fell into complete ruin (Minerva del Cuzco, 1831:40).

By the middle of the nineteenth century, agriculture had been re-established and had become reasonably prosperous again. The lower valley was again producing and the yearly production estimate for Hacienda Santa Ana was 25,000 *arrobas* (1 arroba = 10.4 kgs) of coca; 12,000 arrobas of rum; 100 arrobas of coffee; 300 arrobas of cacao and 250 arrobas of crystalline sugar (PAS SOLDÂN, 1862:399). In 1899 the Peruvian government decided to construct a mule road down the valley through the Vilcanota gorge. Whereas previously the trip from Santa Ana to Cuzco had taken 5 days, it then was shortened to 4 days. Apparently some sort of a road had already existed for part of the way, for in 1880 a sawmill had been established near the present town of Machu Picchu pueblo to exploit the rich timber of the area (GOHRING, 1934:32). Still the route through the valley was long, and if transportation had not been so difficult and time consuming, the commercial hacienda system in the lower valley would have developed more vigorously than it did. Many estates remained only partially commercial because of the high cost of taking products by mule to market. Sugar and coca were favored over other crops for a long time because they could sustain the high freight cost and taxes.

By the latter half of the nineteenth century the valley had recovered from the setbacks of the revolutionary period and the population rose significantly for the first time since the Spaniards' arrival. It is likely that in 1900 at least 100,000 people lived in the Vilcanota Valley. In the first half of the twentieth century, transportation improved greatly. In 1907 some British engineers extended the Southern Railway (Ferrocarriles del Sur) to Checcacupe; in the following decade the line was completed to Cuzco. This standard gauge railway connecting the coast with the Southern Highlands oriented many towns to the railroad and drastically improved the speed and ease of movement in the valley. In the 1920's, the old mule road between Ollantaytambo and La Raya was widened and graded for motor vehicles. Meanwhile, a narrow-gauge railroad was constructed to connect Cuzco with the lower valley and was financed from taxes on coca and rum. In 1927 the line was completed as far as the Machu Picchu ruins, and in 1951 it was extended to Santa Teresa (Huadquiña). Work is in progress on a further extension to Quillabamba. Vehicular traffic in La Convención dates only from 1933, and not until the mid-1960's was there a highway connection with Cuzco.

Population in the first half of the twentieth century rapidly increased, primarily due to a lower death rate. In the immediate basin of the Vilcanota River – not counting the tributary valleys – there was in 1970 an estimated total population of approximately 200,000, a 10 percent increase from the 1961 national census figures (Peru, 1966). The 1961 population for only the Vilcanota Valley of the various provinces were: Canchis, 65,000; Quispicanchis, 25,000; Paucartambo, 3,000; Calca, 23,000; Urubamba, 17,000; and La Convención, 40,000. An estimated 90 percent of these people work directly on the land, although many may live in sizeable towns. The largest towns are Sicuani with about 15,000 inhabitants; Urubamba with about 8,000; and Calca and Urcos with about 5,000 each. In the lower valley the two largest centers are Quillabamba (in 1940 it had 200 people) with about 12,000 and Santa Teresa with about 4,000. Beginning about 1925, an intense migratory movement from the high parts of Cuzco and Apurímac Departments started towards La Convención because of the growing scarcity of resources in the Sierra. The dramatic population increase in the lower valley has occurred despite several malaria epidemics in the 1930's that wiped out approximately a fifth of the people below 1,500 m (CHIRIBOGA, 1937:9). At that time many peasants fled back to the Sierra and the migratory trend was not reversed until after World War II.

PRESENT CHARACTER OF THE POPULATION

The population of the Vilcanota Valley represents a racial, cultural and economic continuum from Indian to white, from indigenous to 'Western' and from very poor to wealthy. Four socio-economic classes may be defined for the valley which tend to follow racial lines. The predominant group is made up of native Indians who may or may not have an admixture of white blood. In Peru, 'Indian' is not defined strictly in racial terms, but rather, more as a socio-economic designation for folk who are not participants in the national society. Indians have certain cultural characteristics in common, and a white man who had these traits would also be considered to be Indian even if he had red hair and blue eyes. With few exceptions, these Indian cultural characteristics are homespun clothes, lack of shoes, patriarchal system and monolingualism.[9]

Indians are extremely conservative and retain many indigenous customs and attitudes, but it would be erroneous to designate their material and non-material culture as 'native', for the colonial Spanish element in their way of life is remarkably strong. More Indians live in the Quispicanchis and Canchis sections of the valley than elsewhere. The Indians seen in the Calca and Urubamba areas actually live above the rim of the valley and come to the valley mainly to trade or for a fiesta.

A transformed Indian class in the valley is the *cholo* group consisting of people who stand culturally – but not necessarily racially – between Indians and mestizos. Cholos may be identified by their factory-made clothing, bilingualism, matriarchal system, preference for 'wayno' music and the white derby hats worn by the women. Cholos dominate the population of the valley and are especially numerous below Calca. Yucay, Urubamba and Quillabamba are cholo towns. In La Convención, where 81 percent of the population is classified as Indian, the great majority have actually been 'choloized'. Whereas all Indians are farmers, many cholos are today artisans and do not till the land.

The third class is comprised of the mestizos who manifest many of the values of Western civilization. However, mestizos do not have a purely Hispanic culture any more than the Indians have a pure Inca culture. They do not perform manual work and are always literate. In southern Highland Peru the more educated among them are called 'decentes' (literally, 'decent people'). Mestizos are often land owners or they may be administrators, teachers or merchants in the towns or foremen ('mayordomos') on manor estates. Some of the larger towns in the valley have considerable numbers of mestizos, e.g. Urubamba, Sicuani and Quillabamba.

9 Approximately 95 percent of the Vilcanota Valley inhabitants speak Quechua, numerically the most important of the South American Indian languages. Roughly 65 percent of these people can also speak and/or understand everyday Spanish, and perhaps 15 percent can also read Spanish. No monolingual Quechua speaker can read Quechua for it is not a literary tongue; all Peruvian schools (are supposed to) teach in Spanish.

Quechua is a highly inflected language and has many gutteral sounds. Spanish has noticeably influenced Quechua and many common Spanish terms are now used when speaking Quechua even when Quechua equivalents exist. Quechua was not a written language until the late sixteenth century and because every writer had his own idea of how to transcribe the phonetics of the Quechua language, no uniform orthography was established. The new orthography devised by Quechua experts in 1954 has not been universally accepted as subsequent writings show, and orthographical anarchy continues. This study will follow the new orthography, except with place names and other terms which are so standardized that to devise new spellings would be improper.

The last class is made up of the whites ('mistis'), people who have no apparent Indian blood and who are the proprietors of many large estates in the valley. Most of them live in the cities of Cuzco and/or Lima (in a few cases, in Arequipa), and only infrequently do they live on their estates. This group numerically forms the smallest class. The vast majority of the Vilcanota inhabitants, perhaps 94 percent, are peasants, either Indians or cholos, who are neither 'primitive' in the ordinary sense of that word, nor are they 'civilized' in the sense of being integrated into modern industrial society.

Synopsis of Present Agricultural Structure, Rationale, Technique and Practice

Agriculture is the main human activity in the Vilcanota Valley, and probably has been so for at least 1,000 years. Both the character of the cultivation in the valley and the plants that are cultivated help explain, and in some measure are explained by, the land tenure arrangements, the technology that has evolved or has been introduced, the economics of land use, the market exchange and adjustments to pests and supernatural forces, as well as by the population characteristics already discussed. All of these factors, in turn, reflect either indigenous and/or Spanish colonial traditions. Some understanding of these factors is therefore essential to grasp the context within which agricultural plants have been and are grown, used and regarded by the people who cultivate them. Certainly cultivated plants do not occur in a vacuum separate from the agriculture of which they form the basis.

LAND HOLDING PATTERNS

The peasant farmers in the Vilcanota Valley today live under three different land tenure systems: manor estate, native community (*ayllu*) or private farm. Essentially two kinds of manor estates or haciendas are in the valley: the institutional estate and the privately-owned estate. In the former kind, the land is usually rented to individuals, *in toto* or in parcels, by bids. The organization which owns the estate is paid part of the proceeds of the harvested crop; formerly the payment was in kind but now it is commonly in cash. Institutional estates are usually inefficient and bring only a fraction of the returns of which they are capable.

A fairly typical arrangement is that of Hacienda Sañahuasi near Pisac, owned by the Dominican monastery in Cuzco. Mestizos rent the three flat areas suitable for cultivation, but the actual farming is peformed by Indians who work one day a week in these lower fields in exchange for land on the unirrigated upper slopes of the estate on which they grow their own subsistence crops. Parish churches also own land in order to support themselves; these glebe lands are usually smaller than those owned by religious orders. In Quiquijana, for example, the parish church owns two small parcels which are rented out to provide income for the parish priest as well as to pay for the annual fiesta. Educational institutions also own estates in the valley; for example, Hacienda Phiri is owned by the Colegio de Educandas del Cuzco (a girls' secondary school), and Hacienda Pachar is owned by the Colegio Nacional de Ciencias (a boys' secondary school in Cuzco). Labor is scarce on these institutional estates and more and more of these lands are being planted in small grains, especially brewing barley, which can be planted and harvested with rented machinery. Most institutional estates in the valley are programmed to disappear in the 1970's under provisions of the agrarian reform law.

Individually owned estates are usually owned by a single person and transferred to the descendants according to the Peruvian hereditary laws. The owner may or may not live on the estate and even if he does, he is absent much of time. All haciendas have a manager ('mayordomo') who organizes the day-to-day affairs of the estate. If the owner is a widow, the manager often makes most of the major decisions concerning the estate. The agricultural laborers ('colonos') on manor estates are those who 'belong' to it and are, in fact, not really free agents. In return for work on the hacienda, they are allowed to farm two or three *topos* for their own use.[10] The right to use these lands passes from father to son. Sometimes this exchange is modified whereby, in return for tools, seed and technical advice, as well as the use of a parcel of land, the peasant is obliged to give the owner of the land a portion of his production, keeping only part of it for his own sustenance.

Because there have been more people in the Southern Sierra region than adequate parcels of available land, may landless peasants were forced to accept the *arrendire* system. In this land tenure arrangement, which predominated in the lower valley until the late 1960's, peasants had the right to use parcels of land in exchange for a certain number of days of work on the estate. In the past, the renter (*arrendire*) worked 180 days a year on an hacienda, plus extra days on such special tasks as clearing irrigation ditches. The essential difference between the arrendire below 1,800 m and the colono above 1,800 m is that the latter is obligated to the hacienda by tradition whereas the arrendire is free to leave as soon as his term of contract expires. The short term contract of the arrendire system, from 4 to 10 years, often precluded the growing of tree crops such as coffee, which produces only after 4 years, and cacao, which begins to yield after 5 years. Instead coca, although often less lucrative, has been planted because it begins to yield in 2 years. In most cases, however, arrendire contracts are extended three or four times so

10 The basic unit of land measurement used since Inca times is the topo. The exact area of a topo has varied from time to time and one cannot be certain of its exact dimensions unless these are explicitly stated. Apparently originally a topo was the plot of land required to maintain one family and thus the actual dimensions varied with the size of the family. In the distribution of lands in Urubamba in 1595 (ANONYMOUS, 1957), a topo measured 100 *varas* by 70 varas (vara = yard).

Today in the Vilcanota Valley estate managers and agronomists use the term topo to refer to an area 84 meters by 42 meters (3,528 square meters). One topo is equivalent to about 1/3 hectare.

Other land measurement units in the Vilcanota Valley which have variable meaning include *fanegada,* the amount of land that can be sown with a *fanega* of seed 1 fanega = $10\frac{1}{2}$ kg.); and *tablón*, a rectangular piece of land usually on a man-made terrace.

that the rented farms are actually quite permanent fixtures. The size of plots varies considerably. Many arrendires become sufficiently affluent to sub-let parts of their rented lands to other peasants who are known as *allegados*. Many arrendires, however, have less than $\frac{1}{3}$ hectare of land on which to sustain themselves and their families, and usually this land is on steep, unirrigated slopes with poor stony soil. Because of government action the arrendire system no longer flourishes in the lower valley, and the laborers are now largely *habilitados* or men who are paid for their work in cash under a 3-month contract but to whom no land is rented.

Throughout the valley, the privately owned haciendas vary greatly in size and in efficiency of operation. The largest hacienda in the valley, Yanahuara, near the town of Urubamba, has 193,000 hectares of land; 80 percent of this area, however, is mountainous and is suitable only as poor-quality natural pasture. Similarly large proportions of the land of the estates in other parts of the valley are occupied by slopes of limited value and usability, and the actual operational size of an estate cannot be realistically determined by its total hectarage. In addition, property surveys are rare, many of the boundary lines are uncertain, and many estates are constantly in litigation with neighboring haciendas or Indian communities. Very few estates are mechanized and none can be said to fully use modern, industrialized agricultural practices. But even considering their modest technological level, the haciendas display a wide range in the efficiency in their use of land. The most progressive units and those with the highest yields are in the Calca-Yucay-Urubamba areas. Virtually all of the fertile bottom land between 3,000 m and 2,400 m is owned by estates. Above 3,000 m there are many large estates and below 1,500 m all land is effectively in the hands of large landowners, although recent efforts toward agrarian reform will eventually eliminate the hacienda there. In 1964, a Peruvian law theoretically abolished many aspects of the feudal land tenure in the country. In 1968 another land reform law was enacted in Peru which promised to radically alter land holding practices; in 1970 several haciendas in the Vilcanota Valley had been purchased by the agrarian reform agency.

Other peasants are members of native communities (*ayllus* or *comunidades indígenas*) which had their origin in Inca times and which have survived in a modified form until the present. The settlement pattern in an *ayllu* is best described as diffusely clustered; individual adobe houses are surrounded by *canchones* or small fenced-in plots of irrigated land. Ayllus, some occupying good bottom land in the valley, are most common in the southern part of Canchis and in southern Quispicanchis. Some of the ayllus are officially recognized by the Peruvian government, others are not. In neither case, however, is farm land still held in common; it now belongs to individuals who pass it on to their offspring, but who cannot sell it outside the community. Unirrigated agricultural land, while no longer owned in common, has its use fixed by tradition; this involves a 7-10 year rotation of land and a watchman from the community who is responsible for everyone's crop (GADE, 1970c). In the fallow period, these lands are used by all for grazing. If planted eucalyptus groves and brush occupy the hills, any member of the community may collect firewood from them.

A fairly typical ayllu is Moccoraise around Cusipata in Quispicanchis. In 1960,

207 families (951 people) farmed 27.3 hectares of irrigated land. The majority of the families had less than $\frac{1}{3}$ hectare of irrigated land; however, several families had none and several had more than 2 hectares. The hillsides were farmed collectively, planted in wheat or barley, and the harvest was divided among the members. Thus the ayllu of today does not ensure equality, but some features of the Inca system remain.

Still other peasants, from 15 to 20 percent of all the valley agriculturalists in the late 1960's, are connected neither with any ayllu nor a manor estate. These farmers own outright small parcels of land varying in size from $\frac{1}{4}$ hectare up to 20 hectares. A property larger than 10 hectares is often referred to as a *quinta*; smaller than that, a *chacra*. Often, but not always, these farmers have some European blood but are culturally cholo. They may have their dwellings on their properties or they may live in a nearby town. Small owner-farmers are found in the valley above 2,800 m but are most common in the Tinta area of Canchis and in the Provinces of Quispicanchis and Calca on the lower slopes of the valley. The lower valley has had the lowest proportion of landowning peasants. In the early 1970's significant shifts in philosophy about the land, fostered by the central government in Lima, started to modify land tenure patterns in the Department of Cuzco including the Vilcanota Valley.

In an area of such ancient and intensive occupance of the land, almost every parcel, no matter how small, has a name to identify it. Property is identified by the names of the individual parcels and not by surveyors' measurements which are in most cases non-existent. Every field (*terreno*) and every enclosure (*canchón*) has a name, many of them plant names, reflecting their immediate natural environment.[11]

AGRICULTURAL LAND USE

The southernmost and highest part of the valley, just north of Occobamba (3,820 m), is the only section in which animal husbandry is more important than crop raising (Fig. 13). Eight large manor estates, together raising thousands of alpacas and sheep, exploit the natural pasture grasses of the slopes of the valley and its tributaries (Fig. 16). Indian shepherds look after most of the animals and bring them to the hacienda headquarters only at shearing time. Down valley from 3,800 m, crop fields of small grains and tubers form a checkerboard in the valley bottom but most of the upper slopes are left uncultivated. Continuing north in the valley toward Sicuani (about 3,500 m) agriculture based on small grains and tubers

[11] In a Quechua place name the prefix of the word is the name of the object; the suffix means the 'place of'. Thus in the valley below 1,500 m there are parcels with the following plant names: Lucmayoc (place of the lucma tree); Paltaychayoc (place of the avocado tree); Piñalpata (pineapple terrace); and Sarapampa (plain of maize). In the valley above 2,800 m one finds a different set of names corresponding to the plants that grow in that climatic realm: Membrilloyoc (quince); Habaspata (broadbean); Tunaspata (prickly pear fruit); Sauceyoc (willow); Molleyoc (molle); Lambraniyoc (alder); Huarnhuinoyoc (waranway).

Fig. 16. Livestock, mainly alpacas, grazing on the native grasses of the puna, elevation 4,150 m.

still predominates on the valley floor, while hardy root crops are grown on some of the slopes (Fig. 17). The damaging effect of temperature inversion on crops planted on the valley floor has not led to a land-use pattern with this factor in mind. Farther down the valley, in Quispicanchis, Calca and Urübamba (from 3,300 m to 2,600 m), irrigated maize dominates the valley floor and wheat is the most important crop on the upper slopes. Generalizing for the whole upper valley section, about 30 percent of the upper slope lands is in crop fields, while the remainder is natural pasture for livestock or too steep (more than 60°) to be usable. In any one year, however, only half of the agricultural land on the slopes is planted; the other half is left fallow. Many of these cultivated slopes have an inclination of as much as 50° and are 600 meters above the floor of the valley.

In few localities is the flattish valley land not used for agriculture. In some places the flat land is mostly infertile, very porous river gravels (e.g. between Urcos and Andahuaylillas); in others it is salt encrusted (e.g. between San Pedro and San Pablo); and in others (as around Racche) poor drainage has precluded cropping. Irrigable lands in the valley floor are intensively used. A field is sown every year without rest, maintained by fertilization and crop rotation. Because of the low night temperatures in July and August and the limited supply of water and fertilizers, the fields generally are not used for crops during the dry season. Where sufficient water and fertilizers are available, however, certain hardy vegetables, such as onions and cabbages, as well as forage barley (*alcacer*) are planted after the

Fig. 17. Ripening fields of wheat and barley on the Vilcanota floodplain south of Sicuani, elevation 3,575 m.

main crop has been harvested. Food production could be increased considerably if low-sun agriculture were practiced on a large scale.

In the ceja de la montaña (2,400 m to 1,500 m), man's agricultural activities are hampered by the absence of suitable sites for growing crops. Most of the valley sides in this area rise sharply at angles of 70-90°, and only occasionally is there enough land for even a small field. A ridiculously small patch of land is given the name of *pampa* in this zone, a term elsewhere reserved for a flat area of several square kilometers. In the valley below 1,500 m, floodplains, river terraces and gentle valley sides (10-20°) provide more suitable places for growing crops than in the previous zone. Unlike the rest of the Vilcanota Valley, La Convención specializes in perennial tree crops and it is this fact, rather than precipitation totals, that gives the landscape a green appearance all year.

AGRICULTURAL TRADITION

Indigenous and seventeenth century Spanish farming techniques dominate agriculture in the Vilcanota Valley. In a few instances, as on the puna, Indians cling to the ways of their ancestors; in others, the European tradition of farming has been almost fully accepted. Mostly, however, the agriculturalists in the valley employ both traditions and the attitudes derived from them.

Perhaps the most fundamental difference between the agricultural traditions of the Old World and those of the New is in their basic attitude towards the crop plant itself. In Inca farming, each crop plant received individual attention with the hoe. The Spaniards introduced a kind of agriculture in which the basic tool, the ox-pulled plow, was used to cultivate the land for a group of plants. This technological introduction apparently was followed by a cultural time lag: the sixteenth-century chronicler ZÁRATE (1968:44) noted that Indians planted wheat in holes made with a stick. Today the general pattern is for indigenous crops to continue to be farmed using the hoe-like instrument, and Old World crops (except for 'garden' plants) to be farmed with the plow and oxen. Wheat is grown, for example, in the Old World manner even on steep slopes, while quinoa is still regarded more as a hoe crop than as a field crop. The cultivation of certain crops may involve both traditions: for example, maize is planted with the aid of plow and oxen, but subsequent work is done using hoe techniques. The broad bean (*Vicia faba*), introduced from Spain in the sixteenth century, is cultivated in either a horticultural or a field manner, depending on the farmer.

Traditional Indian agriculture is oriented more toward security than efficiency. Crop diversity and varietal richness prevent total crop failure. If one member of the plant assemblage fails to yield, other crops – having different requirements and tolerances – will probably succeed. Also the natives seem to understand that different crops extract different soil nutrients. Row crops are intertilled with other crops. Maize, for example, is often interplanted with kidney beans, peas, quinoa, or broad beans. A typical orchard will contain a wide range of useful trees and shrubs. This mixture of many plants in a small area makes harvesting more time consuming, of course, as one must select which plants to remove since they do not all ripen at the same time. The peasants have accepted many Old World crops in their system of polyculture. What appears from a distance to be a homogeneous field often turns out to be a field of highly mixed wheat varieties or a field of barley mixed with wheat. Even more than fields, dooryard gardens are particularly botanically complex, especially at lower elevations.

CULTIVATIONAL PROCEDURES AND PRACTICES

As in most cultures, Vilcanota peasant agriculture is a patterned activity following certain socially accepted methods and techniques of food production with no directives other than those established by tradition. The agricultural work pattern retains some of its pre-Conquest character in the *ayne,* that is, collective activity of a peasant helping his neighbor plow, plant or harvest, on the expectation of receiving reciprocity when he needs similar assistance. The ayne also involves an exchange of a share of the harvest. On haciendas collective work is of a non-reciprocal kind; known as *faena*, it includes the work of sowing, harvesting and cleaning irrigation ditches and extends to non-agricultural tasks such as repairing roads.

The daily work schedule in agricultural activities varies greatly with the season.

During periods of moderate intensity, activity generally starts at 6 a.m., with a rest period at 10 a.m. to drink chicha and chew coca, and a dinner, usually brought to the fields by the women, anytime between 11 a.m. and 2 p.m. At 3 p.m. or so, there is another chicha and coca break and at sunset the workers return to their huts.

TECHNOLOGY

The tools used in growing crops and processing food in the Sierra reflect indigenous and Mediterranean traditions. These tools are inefficient by standards of modern Western agriculture, but technological advances are regarded with caution and often rejected. Some innovations have never been introduced to the valley. For example, the scythe, which cuts grain four times faster than a sickle, is not known, nor is the true flail which can thresh more efficiently than the simple pole now used. Other rejection is a matter of excessive cost. A harvesting-threshing machine, which can accomplish the same amount of work as a sickle in 1/100th of the time, is economically beyond the reach of a typical Vilcanota farmer even if he had three lifetimes. Still other technological rejection is a matter mostly of the kind of job a new tool can perform. On many farms a tractor is impractical because of the steep slopes which it could not climb, and such a machine would not be adopted even if the peasant could afford it. Yet twentieth century agricultural technology has appeared on manor estates, and the capabilities of machines are being more widely known.

LAND CLEARING

Clearing the land of natural vegetation, *wactay*, is a necessary prelude to planting in the forested section of the valley. In the valley above 2,450 m, the climax vegetation has long since been removed and the task of clearing involves only the removal of shrubs and weedy herbs. In the valley below 2,450 m, however, clearing becomes a major preliminary task to agriculture and five traditionalized stages must be completed before crops can be planted: a) the brush is cleared; b) tree trunks are felled; c) debris dries; d) debris is burned and removed; and e) field is burned again. The first two phases of clearing and cutting, called *uhuncheo* in the lower valley, is done in May, June and July. The underbrush – small trees under 6 centimeters in diameter, shrubs, vines, ferns, etc. – is cut with an instrument (*wactana*) which has a steel blade 20-25 centimeters long and 6 centimeters wide. This work gives accessibility to the larger trees and enough elbow room to girdle them with an ax. The cut trunks are allowed to dry from one to several weeks before they are burned. Care is taken to clear a perimeter so that the fire does not spread into adjacent crop fields. Smoke and haze fill the lower valley in August, particularly during the first 15 days of the month when rain does not ordinarily fall. Very often a second burning (*mank'opeo*) is necessary in order to destroy the larger trunks by lighting fires directly on the stumps. By late August or early September rocks and the charred trunks have been removed and all clearing activity is usually completed by this time. Such elaborate procedures are warranted because the land will be used on a semi-permanent basis, rather than abandoned after 3 or 4 years as is done in much of tropical America.

LAND PREPARATION

Whereas land clearing is an agricultural practice mainly of the lower valley, land preparation is of primary importance in the upper valley. Before the field is plowed, the dry stubble and weeds are burned off, thus adding minerals to the soil and perhaps also decreasing insect pests. Often the year's accumulation of hearth ashes is also put on the fields. On the more valuable land, the fields are then irrigated in order to soften the ground as well as to spread the ashes more evenly.

September is the month for most spading or plowing above 2,500 m in the valley. Turning land is still largely done by hand or animal labor as are most agricultural tasks in this part of South America. Peasants have no machinery and even few of the hacendados who could afford it, possess any. As long as a readily available and cheap supply of workers (*peones*) exists, the large estates will continue to use it. In some areas, however, hand labor is scarce or peasant unions have become powerful. Beginning in 1956 tractors have been introduced, but other machines, such as harvesters and threshers, remain rarities in the valley.

The most primitive plow intrument is the indigenous *chakitaklla* or Andean foot plow (Fig. 18). The shaft, usually made of *kishwar* wood in the valley, has the form of an inverted 'J,' and a 10 centimeter-long share is tied to it with leather cord (GADE & RIOS, 1972). In Inca times this share was made of copper or simple fire-hardened wood, but now it is of iron. A small wooden piece, 1/3 meter from the base, functions as a foot lever. The use of the chakitaklla demands teamwork. Traditionally, three men, standing side by side and working backwards up (rather than along) a slope, each opens a furrow with his foot plow. The man in the center (*chaupi*) set the pace for both the man on the left (*k'awaya*) and the man on the right (*k'ollana*). The clods of earth are broken up by two women (*rapa*) who crouch on either side of the central furrow. Today in the valley the chakitaklla is used primarily to break up the grassy turf of steep hillsides inaccessible to plow and oxen. Occasionally flat land that contains many large boulders is worked with this tool. For tuberous crops the land is ridged into a waffle pattern so that good drainage is assured.

The Spaniards brought the Mediterranean scratch plow to the New World and this is the only type of Old World plow found today in the Vilcanota Valley (Fig. 18). This tool, made of *waranway* or *chachacomo* wood, consists of a base to which is tied the metal share. The base is lashed to the beam and a vertical peg runs through both parts to control the depth of the furrow. Branches of shrubs are sometimes attached to the plow to make the furrows wider. Furrows usually run up and down rather than across the slope on the contour. Although such a practice

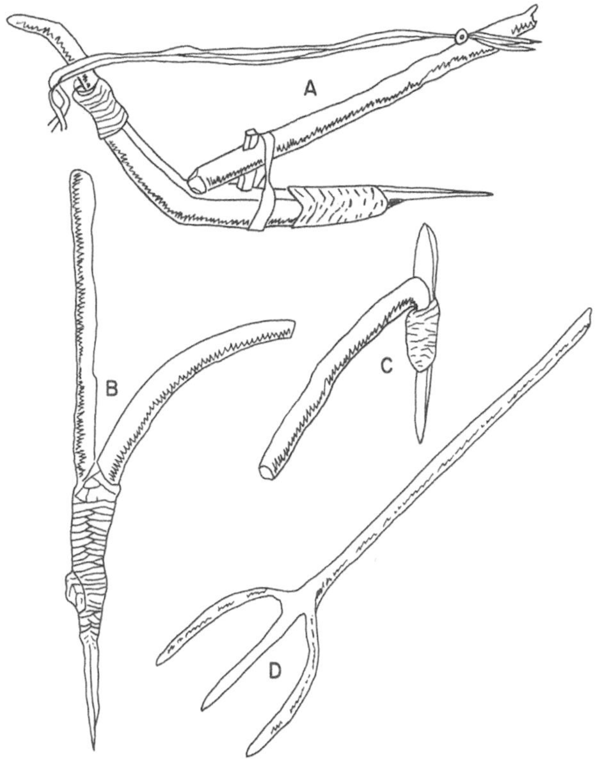

FOUR AGRICULTURAL IMPLEMENTS USED IN THE VILCANOTA VALLEY OF PERU
A. scratch plow (arma); B. foot plow (chakitaklla);
C. mattock (raukana); D. pitch fork (hullka)

Fig. 18. Four Agricultural Implements Used in the Vilcanota Valley.

invites erosion, the peasants believe that the water should drain off quickly lest the crops rot in stagnating water.

This plow (called *arma* by many Indians) is still found in southern Spain (FOSTER, 1960:54) but there the oxen once used to pull the plow have been replaced by mules. In the Vilcanota Valley, the ox is the only animal ever used for this purpose (Fig. 19). The native llama and alpaca were never used as draft animals and for this reason Andean agriculture remained in the hoe stage until the arrival of the Spaniards (GADE, 1969a). The Indians' acceptance of the Old World plow may have been initiated by an edict from Viceroy Toledo (LEVILLIER, 1925:VIII:333) which stated that, in those places where land could be so worked, each Indian community was to acquire a plow with oxen for all its members to use.

Another indigenous tool besides the chakitaklla used in field preparation is the

Fig. 19. Preparing a field for the maize sowing near Ollantaytambo, 2,750 m.

k'asuna, with which one beats the clods to break them up after the land has been plowed. This is a crusher of *lloke* wood or sometimes a round stone 10-15 centimeters in diameter tied to a shaft. Although the clod-breaking harrow was present in Spain in the sixteenth century and may have been taken to Peru, its cost perhaps put it beyond the reach of peasants and so was not adopted.

SOWING

From the middle of September to the end of October, most of the valley farmers between 3,900 m and 2,500 m are engaged in sowing the fields. Both New and Old World sowing methods are used, and these differ basically from one another. In the native Andean agriculture, the individual propagule (i.e. plant part used as 'seed') is deposited in the hole and covered with earth. In European and Near East grain agriculture, the seed is sown broadcast and as such is a much more casual method.

Despite the long tradition of vegetative reproduction in South American root crop cultivation, when they can choose, the peasants prefer to plant seeds rather than cuttings. Most crops that have viable seeds are propagated by seeds. Introduced pome or stone fruits, coca, coffee and cacao are reproduced by seeds even though the peasants know that vegetative reproduction, using slips, permits greater varietal control. Most tuberous plants do not produce fertile seeds and thus the tuber or part of the rootstock, whichever the case may be, is planted. Although planting a large tuber yields a larger harvest, the peasants usually prefer to plant small tubers to conserve food. Fear of theft of the planted tuber soon after sowing is, in some localities, a motivating factor; if stolen, the loss of a small tuber is less keenly felt than if a choice one were so removed.

In the lower valley perennial crops are more important than in the upper valley and sowing follows a somewhat different schedule. After the land has been cleared, food crops such as flint maize, taro, manioc, uncucha and peanuts are sown. Perennial crops such as coffee, coca, or cacao are planted after the maize is harvested; shade trees (such as *Inga*) are also planted at this time. The land thus yields food immediately and supports the peasants until the perennial crops begin to yield.

WEEDING AND BANKING

When weeding, the Vilcanota agriculturalists bank up soil around the roots of growing plants to give them greater support and to make channels for the proper flow of irrigation water. This multi-purpose cultivational practice, known as *aporque*, is apparently indigenous for it is deeply ingrained in peasant mentality and regarded as a necessary practice for successful yields of certain crops.[12] Strong winds, snowfalls and especially hail may more easily flatten a crop that has not had its base banked, and the practice may also prevent drying and freezing. Aporque occurs mainly on irrigated flat land and is most assiduously done with maize. Tall weeds are eliminated with a sickle, but the removal of short weeds and the banking and channeling are done mainly with the *lampa* or *kuti*. This tool is curved hoe with a wooden haft to which has been fitted a semi-trapezoidal metal plate (*llaucho*). The lampa may have originated in the late colonial period on haciendas which had access to metal. Formerly of copper or bronze, most farmers now use steel shares made in England which they purchase at the Saturday market in Cuzco. Best used for aporque work, the lampa may also serve to harvest tubers and roots.

Another important cultivational tool is the mattock or *raukana* which has a narrow steel blade lashed by thongs to a haft, half the length of the lampa (Fig. 18). This mattock is one of the major technological inventions of the Central Andes, and the most important native tool today in the valley. It is noteworthy that a plow may prepare a field for sowing, but weeding is always done with hand tools.

Perennial shrubs and trees such as coca, coffee, cacao and tea are first weeded about 4 months after planting (i.e. usually transplanting). The first weeding is done

[12] Some agronomists believe that the banking up of the root systems is not necessary; however, they are unanimous in acknowledging the value of forming troughs between rows in order to better distribute irrigation waters. The peasants believe that the roots of growing crops are 'weak' and thus must be covered with earth, a practice which may help the plant especially when strong winds blow.

with a asmall tool known in the lower valley as a *k'ituche*. When the plants are older, the lampa is used to weed, bank and to make irrigation channels. Perennials are also pruned, a practice which probably dates from pre-Conquest times.

IRRIGATION

All of the valley is irrigated at least part of the year excepting the zone between 2,200 m and 1,600 m where moisture is superabundant throughout the year. In both the lower and upper valleys, down valley from 1,600 m and up valley from 2,200 m, irrigation is needed, primarily in the months of September, October and November. Usually, but not always, the rains come in full force during December, and from then until April, irrigation is not ordinarily needed. In April and May most crops are in their final stages of maturing in the upper valley and require little water. The specific periods during the year when field crops are irrigated are as follows: a) in June or July to spread the decomposition of stubble; b) before plowing; c) before planting; d) before the first aporque; e) immediately after the first aporque; f) before the second aporque (if there is a second one); and g) once or twice before the harvest, depending upon the amount of rainfall at this time. After each irrigation is a waiting period during which the fields cannot be worked; the length of the period depends upon the time required for the ground to absorb the water. In very clayey soils, work in the fields has to be delayed some 10 to 12 days. In clayey-sandy soils, the period is 7 to 8 days; in loamy soils, the interval is 4 to 5 days. When the soils are very sandy or gravelly, work can begin the day after irrigation.

Irrigation water in the valley is supplied by diverting the glacier-fed streams which flow down the slopes and eventually into the Vilcanota River. Nowhere is water from the Vilcanota moved up and out of the stream bed and onto the fields as is done in the Old World. Norias or Persian water wheels, necessary in Spanish irrigation, are absent from the valley and the indigenous method prevails completely. Irrigation water is regulated by the villages so that each farmer is allotted a certain time-period of water. Irrigation canals, many inlaid with stones, are well established in the Provinces of Urubamba, Calca and Quispicanchis. Some canals used today in the upper valley date back to the Inca period, but by no means are all of the pre-Conquest irrigation systems still in use. Many agricultural terraces, especially those between 3,000 m and 2,500 m were abandoned long ago and are covered today with brush, not because this land is not needed, but because the irrigation canals have been destroyed (Fig. 20). In Canchis, irrigation is much less common than in other parts of the upper valley, and consequently, this area suffers more severely when drought hits.

FALLOWING

Unless the natural fertility of soils is considered, fertilizers must be added if crop

Fig. 20. Inca-made agricultural terraces, 2,550 m, now abandoned.

yields are to be maintained. Generally, crops are rotated on irrigated lands and fields are rotated on unirrigated lands. The most widespread and preferred method of maintaining soil fertility is fallowing or rotating fields, a practice perhaps used during the Inca period on the sloping valley sides. On these unirrigated (*temporal*) slopes, a group of fields (*suertes*) covering a definite area are rotated so that each field is cultivated once every several years. The period of rest, known as *laymi* may be as short as 3 years or as long as 7, and permits the land to store up nutrients so that its subsequent cultivation will yield reasonably well. The length of the rotation period depends partially upon the pressure on the land to produce food and partly upon the accessibility of the fields to settlement. Around Huambutío, for example, where there is very little flat irrigable land, the lower slopes are cultivated every other year, while the upper slopes, 500 meters away, are cultivated one year in five.

CROP ROTATION

It is not certain whether crop rotation was practiced in Inca times or whether it was an idea first brought to the valley by Europeans. OSBORNE (1952:93) insisted that in pre-Conquest times there was no real system of crop rotation, unconvincingly arguing that land which could grow one crop usually could not grow another. ROWE (1946:266) believes it possible that the yearly distribution of land in the Empire was designed to enforce crop rotation. Whatever the case then, today fallowing is preferred over crop rotation. In the core of the maize region (3,000 m - 2,700 m), little crop rotation occurs, but only because large quantities of fertilizers are used. In Canchis, most small farmers recognize crop rotation as a valuable means of conserving soil nutrients. A common sequence is broad breans one year, potatoes the next and maize the third year. In another rotation, wheat or barley is replaced with broad beans and potatoes are planted in the third year. Five-year rotations (1. potato, 2. broad bean, 3. wheat, 4. broad bean and 5. potato) are less common. In the lower valley, some annuals are rotated but, of course, not perennials.

FERTILIZATION

Fertilizers were probably used in the Vilcanota Valley in pre-Conquest times. Llamas carried nitrogen-rich *guano* (bird excrement) from the small islands off the Peruvian coast to the Sierra. *Guano de isla* is still brought up from the Coast today, but most small farmers cannot afford it and must use instead the manure from their own livestock to fertilize fields. The ashes of certain plants are assiduously collected and used as fertilizer, e.g., in Urubamba, farmers burn the vines and leaves of wild cucurbits and certain other herbaceous plants for their potash content and apply the ashes to fruit trees. Chemical fertilizers are still much too expensive for most peasants to use, but a German firm recently constructed a chemical fertilizer plant 15 kilometers from Cuzco on the assumption that peasants would be able to purchase the fertilizer at least in the near future.

PESTICIDES

Chemical insecticides and weed killers are not widely used in the valley because of their cost; even most hacendados cannot afford to purchase them. The peasants' low crop yields in the Vilcanota Valley can be attributed more to the inability of farmers to purchase fertilizers and pesticides, than to misuse of the land. To be sure, erosion is serious in some areas, but compared with many agriculturalists elsewhere, the Vilcanota peasant treats his land with respect and care.

SHADING

The idea of shading perennial crops from the hot rays of the sun appears to be an indigenous practice. The highly traditionalized pattern of growing coca includes the careful shielding of the seedlings and young plants from the sun's rays. Today cacao, coffee and tea are also grown under shade, but these tree crops are shaded throughout their entire life cycle. Manioc serves as a shade plant for all these plants when they are young. When mature, trees are used; the most widespread tree in the lower valley with this function is *kusillo pakay* (*Inga adenophylla*), a native species.

HARVESTING

The main harvest occurs between early May and late June in the valley above 2,500 m, while in the lower valley, it is more dispersed throughout the year. Drinking and music traditionally accompany the harvest. The harvested product, whether seed or underground part, is usually small in size, partly due to low yields and partly by design; the peasants generally value numbers above size, since at high elevations less time and heat are needed to thoroughly cook a small seed or tuber than a large one. The actual harvested return is poor. For example, while each grain of wheat can theoretically produce a spike with 200 grains, the average yield to the peasant is 6 to 10 grains. Many things account for the low return: some seeds do not germinate at all; others are lost because they are buried too deep in the furrow; some are eaten by birds and insects; and many are destroyed by disease.

Tubers and roots are dug out of the ground with any of a number of instruments. The raukana is the main grubbing tool but the lampa, asuylla (smaller than the raukana) and chakitaklla are also used. The *kañari*, a large cow or sheep horn, is no longer commonly used in the valley for harvest work. A few cholo farmers now use the Old World shovel (*pala*).

Grains and other seed crops are harvested either by pulling the plant up, roots and all, the pre-Conquest method, or by cutting the stalks with a sickle (*ichuna*), the method which was apparently introduced by the Spaniards. Quinoa, and occasionally maize, are still harvested in the former manner, but wheat, barley, broad beans and some commercially grown maize are cut with a sickle. Formerly the sickle was made by blacksmiths; now most of them used in the valley are imported from Austria. Although four times as efficient as a sickle, the scythe is not used, possibly because this tool would excessively shatter the grain thus wasting it. In the lower valley, the *k'ituche*, somewhat similar to a sickle, is used for cutting.

THRESHING

Grains and legumes require a more elaborate harvesting process than do tubers or roots, and the time and labor required for threshing and winnowing grain and legumes makes them more work to grown than tubers. Wheat, barley and broad beans involve the separation of a grain from the glumes or seeds from the pods and these are the main crops threshed by animals' hooves. A flat surface, as on a river terrace, is selected as the site for the threshing floor (*era*) which is approximately 4-5 meters in diameter. The grain or legume is heaped into a pile about 1 $\frac{1}{2}$ meters high in the center of the floor. The animals which trample the crop may be horses, ponies, burros, mules or oxen; never llamas. The Incas never used animals in thresh-

Fig. 21. Chola guiding oxen on a threshing floor near Pisac, 3,005 m. The wheat which is being threshed has come from the hillside fields, not the valley bottom.

ing for this technique is a European introduction.

Oxen, because they are heavier, are considered to be better than equines for threshing; the former, however, are slower, making three revolutions on the floor in a minute, while equines can do four a minute. Threshing animals are often rented, typically at 4 soles (US$.08) per day per animal. A driver, standing in the center, prods the oxen to move around the floor and controls their movement by grasping the tail of the ox on the left (Fig. 21). Two people are needed if equines are used: one person, standing in the middle and holding tight to a rope around the neck of each animal, leads the animals; another person follows behind them, encouraging the beasts with whistles, shouts, a whip and sometimes a slingshot. Since animals normally are not muzzled for this work, they occasionally try to eat some of the crop and must be watched closely. As the crop is pushed to the periphery of the floor by the feet of the animals, another person on the side throws the grain back into the path of the beasts with a pitchfork. The quantity of the material threshed varies with the crop. Eight or ten quintales (1 quintal=100 lbs or 45.5 kgs) of soft wheat can be threshed by three animals in one day; this averages out to be the grain harvested from a half topo in 10 hours.

The indigenous method of threshing seeds from glumes or pods is by flailing or beating them with long poles, somewhat flattened on the striking end. The true flail, however, which has a freely swinging piece at the end, is not used in the valley, and the Inca flail may have been the simple pole still used today. The Incas constructed circular, stone-inlaid threshing floors. Today the flailing method is still used under the following conditions: 1) when the harvest is too small to warrant using animals as with field pea, kidney bean, tarwi, quinoa or kañiwa; 2) when some varieties of grain do not respond to treading by animals and must be threshed by hand; and 3) when unthreshed spikes remain on the floor after being trampled by animals. Flailing is much slower than threshing with animals, and not more than 2 to 4 quintales of wheat can be threshed in one day by this manner.

Two primitive threshing devices, the *tribulum* and the *plostellum*, are absent from the Vilcanota Valley and perhaps from all of the Sierra. Apparently of Roman origin, these devices are one technological step up from threshing by animal hooves (FOSTER 1960:55) and they are still used in Spain. The plostellum, a type of cart, was not found in southern Spain in the early sixteenth century, but the tribulum, a wooden sledge with cutting stones set in the base and drawn by a team of animals, was important at that time. Its failure to diffuse to the New World and the Vilcanota Valley in particular may perhaps be explained by the Highland lack of flint for the cutting stones.

WINNOWING

Winnowing is done after threshing and involves separating the seeds from the chaff by the wind. The windy and dry afternoons of June and July and early August are ideal for winnowing. The threshed grain or seeds are thrown about $1\frac{1}{2}$ meters into the air with a three-pronged pitchfork (*hullka*), a sixteenth century Spanish introduction, but made of native chachacomo wood (Fig. 18). Wind carries the chaff about $\frac{1}{2}$ meter away, while the seeds, the dung of threshing animals and the unthreshed parts – all of which are heavier – fall without being carried by the wind. The dung is later picked out by hand and the unthreshed spikes or pods are swept into a separate pile and threshed by flailing. The chaff and stalks (or dried pods and vines) are then swept up to be used primarily as feed for livestock. If winnowing is done in early June, some straw is saved to be burned in bonfires on the eve of the feast of St. John the Baptist, June 24.

SEED SELECTION

The peasants are casual about plant selection, for they see no particular value in keeping varieties genetically pure. Little effort is made to separate different cultivars (i.e. crop varieties) when they are sown, and the usual practice is to sow them together and harvest them at different times, a practical method as long as hand labor is used. If two or more different cultivars flower simultaneously, cross pollination takes place and genetic contamination results. Such selection as does occur takes place during and after the harvest. Seeds for the next sowing are stored in clay vessels but may include among them new mutants that are planted in the next sowing; this explains part of the varietal diversity found in Indian crop fields.

If destined for sale, tubers and seeds are sorted according to cultivar and size; otherwise, they are often not sorted. During the threshing seed products may be very casually sorted: an old woman removes from the main pile those stalks of grain belonging to other cultivars and places them in separate piles.

GRINDING

The indigenous saddlestone and the introduced millstone turned by a water wheel are used to grind (*kutay*) most of the seeds and tubers grown in the valley. The quern and the revolving donkey mill apparently were never introduced to this part of Latin America. The saddlestone used in the Vilcanota Valley is essentially the same kind described by GARCILASO DE LA VEGA (1960:369) as the grinding tool of the Incas. A half moon-shaped stone (*tunac*) rounded on its grinding edge, is rocked back and forth over the seeds which are placed on a flat stone (*mutka* or *batán*). The saddlestone is used to grind small amounts for immediate domestic use. The millstone, driven by a water wheel, has taken over the main task of grinding large quantities in many small Indian communities (GADE, 1971).

The first water wheels in the valley were undoubtedly built on large manor estates early in the colonial period, and it is difficult to say how long the peasants have had water mills for their own use. Triturated wheat and barley; flour of wheat, barley, broad beans, peas or dehydrated potatoes; and sprouted maize are all products from these rustic stone mills, which can on the average grind 100 kilograms of grain in 8 hours. Water mills in the valley are built on the swift lateral streams which flow into the Vilcanota River; no water wheels are on the main river itself. Most of the mills are in the Provinces of Canchis and Quispicanchis, the main wheat producing area of the valley. Notable concentrations of grist mills are near Combapata, San Pedro, San Pablo, Sicuani and Maranganí. A few towns, namely Urcos and Andahuaylillas, now have mills driven by gasoline motors. Although the millstone is more efficient than a saddlestone, even the millstone cannot grind into flour the very hard kernels of several wheat cultivars. Softer durum wheats, however, can be easily ground with it, which explains the preference for these wheats in certain areas. Electric roller mills, now in Cusipata, Sicuani and Urubamba (as well as in Cuzco outside the valley), can grind any kind of grain, and are changing the pattern of peasant milling at an unassessed rate.

CROP ENEMIES

Growing crops may be partially or totally destroyed by diseases, weeds, insects or by meteorological events. When a crop fails, the peasants are forced to adjust in any of the following ways: eat other, still unripe, crops; collect wild 'famine foods', steal from neighbors; sell animals, if they own any; increase coca consumption to ward off hunger pangs; and even emigrate to other areas. Frost, hail or drought are the greatest threats to crops in the upper valley. Because the people

Fig. 22. Wheat stalks which have been tied up near Sicuani (3,590 m) after a March snowfall flattened the grain to the ground.

live at such a marginal level of existence and have little or no provisional security, these natural occurrences can bring famine and near starvation by destroying the maturing crop plants. When frost or hail is anticipated, peasants resort to lighting bonfires, making offerings and praying. Smoke from the bonfires may be of some practical value in protecting crops against frost. Formerly peasant communities tried to ward off a severe frost by ringing church bells, forming a procession and scourging themselves (TSCHUDI, 1847:257), all in reparation for the sins believed to have invoked the impending disaster. Frost is most feared in January and February; if it comes later, the peasants can use at least some of the crop by eating the still-maturing edible parts. Hail or snow flatten plants to the ground. If the grains are not shattered, some of the crop is salvaged by tying the plants in upright bunches (Fig. 22). Hail is at its worst in April and May and is most devastating at elevations above 3,400 m. Occasionally excessive rains may cause rot, especially in growing tuberous crops.

Partial crop destruction may result from fungus, insects, birds, mammals, weed competition and from human theft. Insects and fungi frequently become a scourge in peasant agriculture because chemical insecticides are far too expensive for these very poor farmers. Some plants are relatively free of insect and fungi attacks, e.g., quinoa, añu, ullucu and manioc; others, such as potato, cacao, broad bean and wheat are plagued with pests. The three major fungus diseases which attack growing grain crops are *roya* (*Puccinia* spp.), *carbón* (*Helminthosporium* spp.), and *pakorma* (*Ustilago* spp.). The potato is plagued by *soyra*(*Phytophthora infestans*), the same fungus that caused crop disaster in Ireland. Occasionally fungus attacks in an area can be so virulent that the farmer will replace one crop with another. In 1926, for example, a fungus epidemic almost completely destroyed broad beans which were then replaced by kidney beans. In the lower valley witches broom fungus attacks coca and cacao.

Among the most destructive insects are the following: *papakuru*, a coleopteran larva which attacks the potato; *sillwi*, a lepidopteran caterpillar which attacks the young shoots of both potato and maize; *illa*, a lepidopteran larva which eats potato leaves; *hutukuru*, a lepidopteran larva which eats first the pistils and then the maturing maize ear. Locusts (*langostas*), especially *Schistocera paranensis* have been known to ravage fields in the valley and to cause acute suffering in their wake, but, fortunately they are not common. Below 1,500 m the *kuki* (*Atta sextens fuscata*), a cutting ant, voraciously consumes the leaves of many crops such as coca, coffee, citrus and cacao, but reportedly not those with a milky juice.

Birds threaten crops in the valley, particularly at sowing time and just before harvest. Scarecrows are used in some places and children and old people often guard the fields against bird invasions. In the valley above 2,600 m, several finch-like birds (one known as *konirostre* with black and white plumage) eat ripe wheat and barley seed. In the lower valley, beginning at 2,400 m, the major ornithic nemesis is the *k'alla* (either *Aratinga mitrata* or *Amoropsittaca andicola*) (CHAPMAN, 1921), a small parrot that descends in flocks on maize. The *paukar* a raucous black bird with a yellow tail, damages cultivated fruit trees in this same zone.

Mammals occasionally become nefarious pests to farmers. Both the wild fox (*atok*) and the domesticated dog (*allko*) eat the ears of maize in the 'milk' stage as they did in Inca times. A maize-eating dog may get one foot tied behind its head thus making it incapable of knocking down the maize plants to reach the ears. Foxes are sometimes trapped in maize fields. Much more troublesome are rodents, particularly in the lower valley. Old World rats were inadvertenly imported along with hydroelectric equipment in the 1950's. Below 2,300 m, the rats have moved out of the dwellings and into the countryside in search of food, and the peasants – having little experience with such noxious pests – have little inclination or any effective means of inhibiting their spread and reproduction. Below 900 m, the native agouti or *siwayru (Dasyprocta variegata)* demolishes root crops such as manioc and sweet potatoes. Some farmers around Chahuares (741 m) grow few root crops because they cannot control the agouti. A larger river bank rodent, the paca (*Dinomys branikii*) also eats root crops, but this animal is much rarer than the agouti and therefore less totally destructive.

Another enemy of crops are weeds because they compete with crops for soil nutrients and for space to grow and, in a few cases, are parasitic on crops. However, because weeds have value in themselves as fodder and even, with some species, as human food, the peasants do not consider them to be totally objectionable and many farmers gently tolerate them in their fields.

Widespread human theft of agricultural products may also reduce the food supply of an individual family (GADE, 1970c). Stealing crops from the fields is a patterned form of aggression used by the peasants against one another in place of physical violence. In spite of rigorous state law against it, crop theft was also apparently known in the Inca period. Thievery causes much anxiety and measures are taken to protect the crops. Around ripening time, a small hut (*chuklla*) is constructed in the field and at night a man or boy, often with a dog, stays in it to guard the crop. In small compact communities, one man is often entrusted with guarding everyone's crop. To further decrease the likelihood of crop theft, the fields are fenced either with adobe walls or piles of thorny branches and planted cactus. Fences also help to keep out livestock.

EXCHANGE ECONOMY

More than 90 percent of the people in the Vilcanota Valley derive their livelihood directly from agriculture. Although every manor estate in the valley sells a certain amount of its agricultural production, these units are often not organized on a truly commercial basis, as are the haciendas in the Peruvian coastal region. Only relatively few estates in the valley carry on vigorous trading activity, and a good part of the land of each produces food solely for the owner or manager, his family and household employees. Spatial isolation, low prices, scarcity of labor in some cases and, most of all, general disorganization force these valley estates to remain only semi-commercial. Crop emphasis may shift. In the 1950's and early 1960's coffee became the prime commercial crop in the lower valley due to high coffee prices, coca's uncertain future, and a high tax on rum. As transportation improves from the lower valley, one can expect a greater attempt to produce perishables such as fruit. Workers have always been scarce in the lower valley and if they had not been perhaps agricultural production would have been more varied. As it is, almost all the commercial crops are perennials which are not overly susceptible to disease (except cacao) or insect plagues, and which do not require fertilizer, an item too expensive to be brought into the area.

Among the peasants themselves, some engage primarily in commercial endeavor, but the vast majority have mainly a subsistence orientation. Few peasant farmers in the valley depend upon a cash income to buy basic food necessities. A close look at the economics of production indicates that the valley peasant earns very little indeed by farming; in fact with the existing land use and organization, he can barely eke out a living. For example, a small farmer in the Province of Calca, renting 2 topos of irrigated land, pays the landlord 120 soles (less than US $3.00) a year. The price of seed is 40 soles; the cost of cultivation and harvest amounts to around 100 soles. He cannot afford fertilizers and thus his yield does not exceed 7 fanegas (1 fanega = 11.4 kgs) of maize per topo. An optimistic yield of 14 fanegas from the 2 topos can be sold at 80 soles per fanega, providing a total income of 1,120 soles. Subtracting 260 soles for expenditures, he retains 860 soles. This amount (which assumes that he sells all and consumes none of this maize) is very little for even the barest necessities of life. Nevertheless, a disproportionately high part of it goes for alcohol which, given the present socio-economic structure of the valley and the responsibility of the peasant to be 'sociable' or 'hospitable', may almost be considered to be a 'necessity'.

Although the agricultural economy of most small peasant farmers in the Vilcanota Valley is basically subsistent in nature, much exchange occurs among farmers. The most active markets are those which receive products from different environmental zones. For example, at Chincheros, a village far above the valley from Huayllabamba, potatoes are the main crop and people from the depression below come to exchange their specialities for this tuber: maize, firewood and walnut leaves are brought from Urquillos; fruit and bread from Urubamba; vegetables from Huayllabamba; and maize and strawberries from Yucay and Ollantaytambo (ESTEVA FABREGAT, 1970). The Tinta market functions mainly to exchange valley grain for animals from the heights above in Canas. Active markets such as one at Sicuani may also develop because many Indian communities tend to specialize in one crop. Thus in the southern part of the valley at least three communities emphasize wheat, three focus on potatoes, and one on onions.

Prices for agricultural products in the native markets adjust to supply and demand. During and shortly after the harvest, prices are considerably lower than, say, around the first of the year. Also, prices differ from market to market; the value depends on the distance from which the products are brought and the natural scarcity of such items in the local area. Thus maize has a higher value in Sicuani than do potatoes, broad beans, barley, wheat or quinoa. But in Urubamba all these latter products cost more than maize which is plentiful here.

Barter, called *waki* in Quechua, continues today wherever Indians are both the buyers and sellers. The value of each product seems indefinitely set but it is obviously recognized by the people themselves, for little verbal confrontation accompanies the transaction. One dried ear of maize may be traded for one ball of wool yarn; a clay pot may be exchanged for as much maize as fits inside the vessel; a live sheep is bartered for 10 kilograms of wheat flour; two bunches of onions go for $\frac{1}{2}$ kilo of dehydrated potato and so on. In markets frequented by mestizos Peruvian currency (*soles oro*) is the common medium of exchange today. This demand for legal tender has helped change the agricultural orientation of some areas. The lower valley has become an area of money exchange because mestizo merchants demand cash in order to purchase grains, legumes and tubers from the upper valley. Crop specialization with resultant cash sales has put the Indian at a disadvantage because he cannot produce crops efficiently. When the price of a particular crop fluctuates greatly from year to year, haciendas no longer cultivate it and the production of the crop is left to the peasants; such has been the case with coca, achiote (a seed dye) and cacao in the valley.

The exchange of products takes place in daily, weekly or yearly markets. Daily markets are set up to trade goods coming from a radius of several kilometers. Many towns in the valley have, in addition to these daily markets, weekly markets at which not only local products are sold, but also those of the entire region. On Sundays, Pisac, Combapata, Quillabamba and Sicuani have busy markets; on Wednesday, Urubamba; on Thursday, Tinta; and on Saturday, San Pablo.

The Pisac Sunday market, a completely outdoor affair, is organized around the exchange of products from the valley above 2,500 m with those from the nearby puna above the rim of the valley. On this day the Indians from the cold lands bring down their dehydrated tuber products and trade them for maize, wheat and cabbages. Some products from the lower valley are also involved, primarily coca which is an indispensable item to Indians. The plant products seen in his market on February 2, 1964, included the following:

ears of dried maize	capsicum pepper
dehydrated potato (*chuño*)	balls of quinoa ash (*llipta*)
new potatoes	pears
fresh peas (in pod)	peaches
wheat kernels	Brassica greens
dried broad beans	dahlia blossoms (ornamental)
dehydrated oca (*k'aya*)	garlic
ullucu (tuber)	cabbages
achokcha (vegetable)	barley
pakay pods (fruit)	tree tomato (*Cyphomandra*)
coca leaves	dried peas
tuna (cactus fruit)	palillo (a root dye)
achiote (a seed dye)	capulí cherries
sprouted maize (*jora*)	sweet lemons
cucurbit (*calabaza*)	Cavendish bananas
lettuce	herb bunches (*asnapa*)

Quillabamba, the market center of the lower valley, receives many products from the upper sierran area along the railroad: potatoes from Izcuchaca; onions, barley, potatoes and ullucus from Huarocondo; maize and cabbages from Ollantaytambo. The Quillabamba food market is both inside a building as well as on the streets around it. Prices are controlled and vendors who exceed the set ceiling may be fined. Space on the street is rented from the municipality for a few cents; inside the building, the rents are much higher. Inside all the vendors spread their goods on tables; outside only some of them do; the majority sit or squat on the street. A characteristic of the outside chola vendor is that she sells a greater variety of products than do the cholas inside. Most of the vendors sell both tropical and temperate food products; the former come from the vendor's own chacra, the latter were purchased in the villages mentioned above and brought down valley by the vendor. Thus one woman may sell potatoes, onions, ullucus and cabbages, as well as cooking bananas, pineapples and papayas. On July 10, 1963, the following plant products were seen in the Quillabamba market:

tangerines	dehydrated potatoes
oranges	ullucus
sweet lemons	camomille
Gros Michel bananas	carrots
pineapples	garlic
passion fruit	capsicum peppers
papayas	broad beans (dried)
avocados	lettuce
manioc	oca
uncucha	tomatoes

sweet potatoes	potatoes
cherry tomatoes	cauliflower
plaintains	apples
red onions	dried flour maize
popcorn	wheat
barley	herb bunches (*asnapa*)

Yearly markets are held in conjunction with religious harvest festivals. The most important one occurs on August 15 in honor of Our Lady of the Assumption ('Nuestra Señora de la Asunta'), held at both Tiobamba (about 10 kilometers west of Urubamba, actually outside the valley proper) and at Calca. Another festival, Nuestro Señor de Huanta, is held near San Salvador in September. In addition to the dancing, drinking and praying that go on at these large fairs, many products are traded: handwoven textiles made by Indians; factory-made clothes sold by and for cholos; pottery and utensils, livestock, mainly burros, mules and horses and formerly llamas; wheat, maize, fruit, palillo, popcorn, tubers, chicha and prepared dishes such as roast guinea pig. The maize comes from the Urubamba area, the tropical fruit from Quillabamba, tubers and chuño from the cold lands above the rim of the valley and from the Altiplano, wheat from Canchis and chicha from Urubamba, Yúcay and Huayllabamba.

MAGICO-RELIGIOUS INTEGRATION WITH AGRICULTURE

Each phase of farming from plowing to harvesting is accompanied by acts of piety which may be pagan, Christian or a mixture of both. Religious and magical beliefs are integrated into the cyclical tasks of livelihood, suggesting a pervasive sacred attitude toward the material universe lost in modern industrial societies as natural processes have been given a scientific basis (GADE, 1970b).

As in many peasant societies, the female principle in Vilcanota agriculture is strong.The earth mother (*Pachamama*) cult has survived more than most indigenous beliefs. This cult had no organized Inca priesthood as did the worship of the sun, so the early Spanish clergy did little to break it up; in fact, the Spaniards even permitted stones and other earth mother fetishes to be placed at the feet of images of the Virgin Mary (MAXWELL, 1956:62). Prayers invoking her aid still accompany the basic agricultural processes. Peasants place elongated boulders in the fields believing that, through the intercession of the earth mother, these stones will conserve the humidity of the land.

In aboriginal thinking, the earth mother is the divine being which animates all life, including many useful plants; thus it has been usual to identify her with certain important crops. Most common is the maize-mother (*sara mama*), the tangible image for which is a stone representation (*konopa*) of a maize ear or an unusual dried ear, either extra large (*choklo kuti*) or double (*takke*). These maize images are kept in a conspicuous place in the hut, often in front of religious pictures, and will not be consumed. Another manifestation of the female principle in agriculture is the prevalence of women as sowers. The man plows, weeds and

Fig. 23. In the sowing of tubers, the man is turning the earth with a chakitaklla while the woman places the propagule into the ground. The other person on the right has a *k'asuna* or clod crusher which is used to break up the large clods.

harvests the crop, but it is inevitably the woman who drops the seed into the ground; such was true also in Inca times (Fig. 23).

Some Indian communities in the valley still entrust to an individual the responsibility of warding off natural disasters. The watchman (*arariwa*) builds a small shrine in the most prominent place in the area and when a hailstorm threatens, he burns incense and piles of grass and buries coca leaves and maize kernels. A sixteenth century drawing by Guaman Poma de Ayala indicates that this is an indigenous ritual (Fig. 25). In a few places in the valley, particularly in Canchis, witchcraft is surreptitiously practised to ensure large crop yields. Money is collected from each farmer to pay the witch and to buy such offerings as candy, coca, cigarettes, *wayruro* beans, alpaca fat, wine, carnation blossoms, cookies, eggs, rum and candles. The witch performs incantations and invokes the ancient Inca spirits as well as Christian saints (especially Santa Barbara and the Virgin Mary) to have compassion and to spare the people from the destructive effects of weather phenomena.

Agricultural superstitions fill the peasant's life, though they are more significant to Indians than to cholos. Many peasant folk believe that bad harvests are a punishment for sin: a drought or hailstorm is regarded as a direct divine intervention by which the people are chastised for their evils. Potatoes and maize are often not sown in fertilized fields during the period of new moon and quarter moon because the crops so cultivated are said to not yield well. Another superstition is the

Fig. 24. Here the Indians are working in the potato harvest. The boy on the left is digging up the plant with a chakitaklla while the girl on her knees is using the *raukana* to complete the task. The person on the right has placed the tubers in a cloth sack and is carrying them to the sorting or storage area.

believe that the green fruit of some plants (such as cucurbits) should not be touched for fear the fruit will consequently rot before it ripens.

Several Spanish religious fiestas of agricultural significance have been transferred to the valley. The feast of St. Isidore the Farmer on May 15 (October 26 in the Northern Hemisphere) has taken on a greater importance and a more specialized nature in the Vilcanota Valley than it has in Spain itself. St. Isidore was a farmer near Madrid who is widely venerated as the patron of peasants in the Spanish speaking world. He was not canonized until 1622, but a cult probably had developed around him by the time of Pizarro and so was exported to the New World. In the Vilcanota Valley this feast day fits well into the harvest period, a time of great anxiety. Tinta, San Pedro, Combapata, Pisac and Calca have fiestas at this time. In Calca, the image of San Isidore is adorned with the best spikes of wheat and is carried in a procession around the town. Behind the saint's image follow several teams of oxen festooned with garlands of flowers. In the plazas of the smaller towns, invocations are made to the earth mother, to the ancestors (*apu*), to the divinities in the hills (*auki*), as well as to San Isidro himself. Libations of rum and chicha always accompany these solemn acts. Although in all of these villages, maize cultivation is more important than that of any other crop, the

Fig. 25. A thief is seen in the act of stealing maize ears, and a llama has trespassed into the field. The watchman on the hill above is supposed to be guarding the crop, but is apparently engrosed in his bonfire, made to propriate the spirits and to ensure a successful harvest. This practice is still known today.

associations here are Old World: wheat, oxen, the plow. No New World crops are part of this agricultural fiesta. During the actual harvest, field ceremonies are held in which a crude cross, fashioned out of grain, is placed on the top of the collected piles of the harvested crop and the leader, hat off, kneels in front of the piles, invoking San Isidro and sprinkling the harvested grain with rum. The harvest cross (*cruz de la cosecha*) may have been a medieval Christian symbol of gratitude for the harvest; today it is used for crops that are both indigenous and introduced.

The feast of the Purification of the Virgin (*Santisima Virgen Purificada*), February 2, occurs when the fruits of crops are beginning to appear, a time of anxiety about hail and frost. On this day, very early in the morning, the people bring firewood and dried cow dung to make bonfires in which are placed incense, llama grease and coca leaves as burnt offerings. As patroness of the harvest, the Virgin Mary is expected to impart her blessing on the growing crops. Some farmers still refuse to work in their fields on certain feast days such as this, a practice probably initiated by an injunction of the Spanish clergy in early colonial times.

Fig. 26. This scene is most likely to occur in the month of February. Early potatoes (*maway*) are being harvested under an overcast sky. Meanwhile a dog, two foxes and some birds are menacing the corn crop which is in the 'milk' stage. A woman stands on the field margin and beats a drum to frighten the animals out of the field. Dogs and foxes are still regarded as threats to maize, and woman and children are still enlisted to frighten pests away. In colonial times a religious rite involved 'excommunicating' pests that endangered growing crops, but this custom has apparently died out, at least in the Vilcanota Valley.

2. TIME, ECOLOGY AND SPACE: THE GEOGRAPHICAL SYNTHESIS

Cultural-Historical Perspective of Useful Plants in the Vilcanota Valley

The relationship between plants and man may be expressed as a continuum from extreme human manipulation of the species to no apparent human intervention (Fig. 27). Viewed in this way, three broad groups of plants emerge: domesticates,

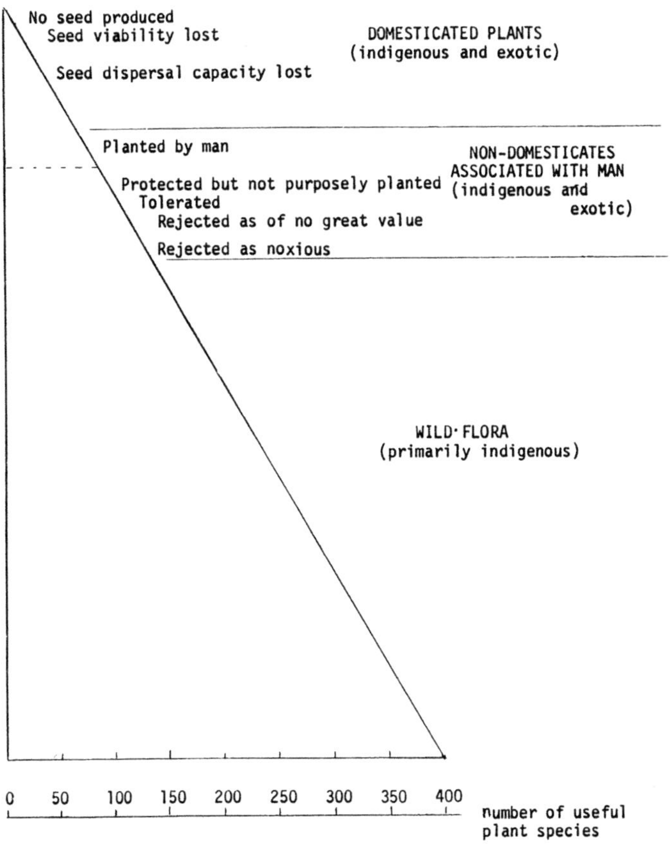

Fig. 27. Typology of Human Manipulation of Plants.

non-domesticates associated with man and the wild native flora. Domesticated plants are set apart from their non-domesticated counterparts because man, by selection, has genetically transformed these living plants making of them an artifact, and it is man who determines their distribution. For any given area, man chooses from among those domesticated plants which can tolerate the conditions set by nature and which can effectively compete with other plants for the available energy and nutrients provided by the environment. The wild plants of any area, on the other hand, are usually members of the native flora that have not been changed by man, and have their distributions determined by seed dispersal capacities, conditions of the natural environment and ability to compete with other wild species. Between these two extremes is yet another group of plants: those which have been favored by the presence of man, often to the point of actual cultivation, but which do not depend on man for propagation or survival.

DOMESTICATES

All domesticated plants are artifacts, but some are more genetically plastic than others, or have been altered by man to a greater degree. Much manipulation is apparent in plants which have lost the capacity to produce any seed at all (such as oca and ullucu); those which have lost the capacity to produce viable seed (such as añu, achira, and pepino); and those which do not have the capacity to disperse their seed (such as maize). Most domesticates are also cultigens, i.e., their wild ancestor is unknown, being the result of folk selection during more than a millenium. Man may be partially at least responsible for the absence of the wild counterpart of many Andean crops: agriculture may have been extended into every piece of land possible for crops, and such spots may have been the only places where the wild ancestor could grow.

Domesticated plants have appeared in the Vilcanota Valley in roughly five geo-chronological sequences: 1. Andean domesticate adoption; 2. pre-Columbian New World introductions from other centers of crop origin; 3. post-Conquest New World introductions from other centers; 4. post-Conquest Old World introductions in the sixteenth century; and 5. recent introductions.

Andean Domesticates

Although evidence for crops is at least as old as 3,000 B.C. on the Peruvian coast, it is unlikely that actual domestication took place here because the arid coastal environment is not the natural habitat for the wild plant relatives that could feasibly be the ancestors of any Andean crop. The idea of domestication in South America may well have been originally derived from forest dwellers to the east of the Andean valleys where root agriculture is now known to be very old. It is reasonable to postulate that these forest farmers slowly moved into the Andes, adapting those root crops which were amenable to domestication and leaving behind those tropical plants which could not adjust to lower temperatures (SAUER, 1952:50). A wide variety of environments in relatively short distances meant that the example of plants domesticated in one habitat could easily be transferred to plants in another habitat. Protein needs may have led to the selection and gradual improvement of those plants with edible seeds. Basically, however, native Andean agriculture is primarily root and tuber cultivation and only secondarily is it based on plants grown for their seeds. This is in marked contrast to agriculture in Highland Mexico, where seed agriculture may have evolved out of a long tradition of collecting wild seeds on the dry plateau of Mesoamerica.

The Andean center of domestication, as defined by VAVILOV and his collaborators (1949-50), extended from Colombia to northwest Argentina. Although Peru is the locational heart of this latitudinal extent, it may not have been the center of Andean domestication. SAUER (1965:305), for example, now leans strongly toward a Colombian center of domestication where an open corridor of rich volcanic soils leads up from the Caribbean Lowlands into the Highlands without brusque environmental changes.

The Vilcanota Valley provided ideal physical and cultural conditions for the domestication of useful plants and their subsequent ennoblement. Sharp temperature fluctuations and high levels of insolation may have induced mutations in crop plants to man's advantage. Irrigation is not usually essential for crop growing in the valley, yet where water is applied to the fertile natural river terraces and bottom lands, yields are even more rewarding than usual. However, present knowledge of crop origins makes it impossible to say which, if any, crops were domesticated in or near the Vilcanota Valley, or any other place of comparable scale.[13] It is less hazardous to posit the valley as the setting for the development and improvement of certain crop varieties of the potato, maize, quinoa and coca, as a result of folk selection over many generations.

The labor-intensive farming practiced on the terraces so painstakingly carried out and built on the steep slopes may have played a part in the development of a certain number of Andean crops. Judging from the complete absence of stones in the soil of many terraces used today, it is possible that this soil was transported there *in toto* from alluvial bottoms. These terrace soils were probably not systematically fertilized before Inca times when llama trains from the Coast carried nitrogen-rich guano to the Sierra. Terraces, many of them still intact, certainly pre-date the Incas, although the Incas expanded and improved the system. On thse platforms each plant received individual care and a farmer could often spot, then select and propagate, mutations or hybrids because gene changes are sometimes visibly manifested in the plant. In these biological modifications, aboriginal man – unlike the modern scientist – simply recognized and selected the chance variations which occurred. Even inferior mutations were preserved by primitive farmers because they did not depend on outside sources of seed for their next sowing.

[13]Although there are in the valley wild species of many genera which include crops (such as: *Solanum* [potato]; *Oxalis* [oca]; *Tropaeolum* [añu]; *Chenopodium* [quinoa and kañiwa]; *Canna* [achira]; *Amaranthus* [kiwicha]), few plausible crop ancestors are among them.

It is highly likely that different local cultivars were thus developed in the Vilcanota Valley. The elaboration of crop varieties was aided by limited human movement and thus minimum mixing of genes. During the Inca period (c.A.D.-1100-1532), agricultural knowledge and technique was refined to the highest levels known in pre-Columbian America, and food production from plants provided large surpluses on a sustained basis. Undoubtedly the Incas were responsible for the ennoblement of at least some of the crop plants on which they depended.

Crop Contributions from Elsewhere in the New World

In addition to those domesticated plants which are fairly clearly of Andean origin, the Vilcanota Valley also has domesticates which originated in other New World areas but which were brought to the Andes in pre-Columbian times. From eastern South America, south of the Amazon, diffused such cultigens as manioc, peanut and the pineapple. From Mesoamaerica came maize, avocado, lima bean, kidney bean and squash (*C. moschata*). These diffusions occurred before the time of Christ; hence since their introduction, sufficient time has elapsed for local cultivars of these plants to be created by folk selection. Indeed, few places display such a diversity of maize varieties as do the southern Highlands of Peru.

After the Conquest, the Spaniards moved products among their colonial territories. Several plants which are important in the Vilcanota Valley today were probably brought in the sixteenth century from Mexico or Central America either by the Spaniards themselves or possibly with the numerous Indians imported as slaves in the 1530's (LOCKHART, 1968:199). These include capulí cherry, maguey (*Agave americana*), tuna cactus (*Opuntia*), papaya, cultivated ornamental dahlia, the condiment wakatay (*Tagetes minuta*) and perhaps the tomato and one variety of cultivated cacao.

Transfer of Plants from the Old World

The plants which have had inestimable importance in changing the cultural plant geography and agriculture of the valley are a group of domesticates which originated in the Old World. Although brought to the New World by Europeans, the plants are mostly of African and Asian origin and not originally from Europe at all. We are certain that Old World plants entered Peru beginning with the Spanish Conquest, but much less sure of any previous trans-oceanic plant diffusion. The archeological record (TOWLE, 1961:93-4) confirms the pre-Spanish presence in Peru of the Old World gourd, but this is the exception. The dearth of substantive evidence of an earlier plant migration, however, should not close the book on the distinct possibility of such an occurrence (GADE, 1970a:207), and further speculation is to be encouraged. It is beyond the scope of this study to examine the suggestions of C.O. SAUER and others that sugar cane, castor bean and plantain were all in pre-Columbian South America. A *de facto* lack of documentary evidence on this topic should not mean that such occurrences did not and could not have taken place.

Beyond speculation is the fact that virtually all of the elements of Spanish agriculture had been successfully transplanted into the Sierra and the Vilcanota Valley by the end of the sixteenth century. This transferral was remarkably rapid, aided by an order issued in 1532 by the *Casa de Contratación* in Seville that required every ship going to the New World to bring seeds, live plants and domesticated animals (WHITAKER, 1929:4). The Spanish government regarded these introductions as part of the task of civilizing these lands. But there was also the more practical desire to make Peru self-sufficient in foodstuffs, a desire prompted by the high shipping rates and the tenuous sea connections between Spain and Peru. Governmental interest may have continued a long time, for as late as 1806, ALEXANDER VON HUMBOLDT (1811,III:149) wrote that no European government was more deeply committed than was Spain in promoting the study of plants and animals, mostly for the New World. The mother country, however, disapproved of the large-scale cultivation of certain commercial crops such as olives, hemp, mulberry, flax and the vine, because these competed with important Spanish exports.

Acculturation of Old World Crops

Plants introduced from the Old World were immediately accepted only by the Spaniards themselves. The indigenous people preferred their own native crops and agricultural techniques to satisfy their own needs and desires. The encomienda system and the demand for tribute greatly speeded up the Indians' crop acculturation (KUBLER 1946:354). With much of the good agricultural land in encomiendas, Old World plants had a much better chance of becoming established than without Spanish control. Tribute was in kind, mainly food products and frequently Old World food crops, and this eventually helped to bring acceptance of introduced cultigens. Later in the seventeenth and eighteenth centuries, much land fell into the hands of the Spaniards who previously only had legal rights to the use of Indian labor. These haciendas then produced Old World plants and animals; for example, at the Jesuit estate near Pisac the main products in 1649 were wheat, broad beans and sheep (ANONYMOUS MS, 1649). Around mission establishments the Spanish clergy also had gardens of European horticulture which helped to disseminate Old World garden crops.

Tribute was still exacted from Indians who worked their own lands; for example, in 1729 the tribute for the town of Taray in the valley amounted to about 45 kilos of maize and 11 kilos of wheat (ANONYMOUS MS, 1729). Although these figures do not reveal the quantity of wheat actually cultivated at Taray, by this time the Indians may have grown some wheat and other Old World crops for their own use. Historical documents collated by KUBLER (1950:342) suggest that after 1750, Peruvian Highland Indians grew European small grains for their own consumption rather than merely as tribute. Thus apparently there was a cultural lag in the Vilcanota Valley of 150-200 years before Old World agriculture was broadly accepted by the Indians. The period of greatest acculturation was the 100-year period before 1750 during which Old World plants were grown on a fairly small scale. It is possible, although difficult to prove, that crop diseases or climatic catastrophes influenced the peasants to finally accept Old World agriculture on a par with their own. In addition, resemblances to known native plants may have facilitated the acceptance of some crops. For example, the flavor of the onion bears some similarity to the edible bulbs of the wild chullkus *(Nothoscordum)* collected by the natives. Likewise the carrot resembles the arracacha in texture and vaguely in flavor.

Considering the New World and Old World crop complexes side by side, it becomes a useless argument to assert the superiority of one over the other. Both were diversified and had the capacity to sustain large sedentary populations. Aside from oil plants, both complexes were nutritionally on much the same level. With the arrival of the Spaniards, the total agricultural inventory of the valley was greatly enriched, and after the indigenous population accepted this new agriculture, the nutritional and economic gains were considerable. Only above 3,900 m elevation did the Old World agriculture not contribute any significantly useful plants. Conversely, the greatest total contributions were made below 1,800 m in the macrothermal sector of the valley where numerous Old World tropical plants found a suitable environment.

In a comparison of the staple foods, perhaps the greatest question revolves around the two basic grains, maize and wheat. If both are grown without irrigation, wheat clearly is a much more satisfactory crop because it yields well in sub-humid conditions. With irrigation, however, maize is superior, for its yields are much higher. The Spaniards probably recognized the quantitative superiority of irrigated maize, even though they may have preferred wheat as a food, and they valued maize on a par with small grains almost from the beginning. In 1658 in the valley around Lamay, wheat was cultivated on the unirrigated hillsides and maize in the more fertile and well-watered valley bottoms (ANONYMOUS MS, 1658). Maize is somewhat more resistant to epidemic disease but it is more liable to frost damage. Nutritionally wheat has more protein but less carbohydrates and fats.[14] Maize can be prepared in a greater variety of ways than can wheat, although only wheat makes satisfactory bread. Considering plant parts other than the seed, the maize plant has a greater usefulness to man than does the wheat plant.

Many of the crop varieties introduced from Spain are still grown in the Sierra and the crop flora in the valley is today more archaic than that of most of Spain itself. Some present crop varieties of Old World introductions may not have their origins in the Old World, but may be the result of folk selections over the past 400 years. However, lacking precise colonial records of plants actually brought to Peru, the establishment of certain varieties as folk selections is impossible to demonstrate.

[14]GIRAL (1966:256-7) points out that the rarest and perhaps the most valuable essential amino acid to humans is tryptophan. The main protein in maize – zein – contains no tryptophan, while the equivalent protein in wheat, gliadin (which forms gluten) contains about 1 percent tryptophan.

Rejection of Plants

Some plants introduced in the sixteenth century have never been as successful in the Vilcanota Valley as they were in Europe. The climatic rhythm existing in the Sierra is not in harmony with the vegetative rhythm of some plants, or, more logically, the hereditary rhythm is not in equilibrium with the environmental rhythm. For example, because there is no well-defined dormant season in the Sierra, grapes grown here are very inferior. Other plants were not successful because their cultivation required too drastic a change from the indigenous way of doing things, or they did not make an efficient use of the land. KUBLER (1946:357) argued that unless its value is so great as to offset the extra expenditures or adjustments, a cultivated plant will not be accepted. He cites fruit trees (apples and pears) as examples, but their relative lack of success today in the valley is more due to the lack of a well-defined dormant season than the reasons he stated. Still other plants were not well accepted because they duplicated, rather than complemented, nutritional uses of the native crop plants. For example, the Indians had a wealth of root crops and saw no need for such a plant as the turnip.

Subsequent Introductions from Old and New Worlds

After the sixteenth century, only a few plants were introduced which became widely adopted. The most important of these were coffee, mango and eucalyptus. In the 400+ years since the Conquest, many plants, though introduced, did not assume agricultural importance in the valley. Beginning in the late nineteenth century, a number of plants were introduced into the valley, but they were destined to remain agricultural curiosities for one reason or another. Zealous agronomists with visions of an agricultural paradise never realized were responsible for some of these ephemeral introductions. Other short-lived novelties were made by ambitious estate owners such as Sr. Pedro Duque of Hacienda Santa Ana who brought in many plants (of both Old and New World origins) from Colombia. The reasons most of these plants were rejected are complex. Cinnamon, first planted in the valley in 1946, failed after ten years of experiments because of cutting ants and unseasonal weather. In most cases, however, cultural or economic reasons counted more than the physical growing environment did. Below is a list of abortive introductions (sixteenth and seventeenth century introductions are starred).

Cereals (or pseudo-cereals)
millet (*Setarea* or *Pennisetum*)
buckwheat (*Fagopyrum sagittatum*)
sorghum (*Sorghum vulgare*)
rye (*Secale cereale*)*
canary grass (*Phalaris*)*

Oil Plants
sunflower (*Helianthus annuus*)
sesame (*Sesamum orientale*)

Legumes
chickpea (*Cicer arietinum*)*
lentil (*Lens esculenta*)*
cowpea (*Vigna* spp.)
pigeon pea (*Cajanus cajan*)

Root Crops
turnip (*Brassica rapa*)*
parsnip (*Pastinaca sativa*)
arrowroot (*Maranta*)

oil palm (*Elaeis guineensis*)

soybean (*Glycine soja*)
olive (*Olea europea*)*

Fleshy Fruits
cashew (*Anacardium occidentale*)
pomegranate (*Punica granatum*)*
mamey (*Mammea americana*)
breadfruit (*Artocarpus altilis*)
sapote (*Casimiroa edulis*)
jujube (*Zyzyphus jujuba*)*
grape (*Vitis vinifera*)*
tamarind (*Tamarindus indica*)*
watermelon (*Citrullus vulgaris*)
date (*Phoenix dactylifera*)
fig (*Ficus carica*)*
carob bean (*Ceratonia siliqua*)*

Industrial crops
Egyptian cotton (*G. barbadense*)
jute (*Cochorus capsularia*)
hops (*Humulus lupulus*)
fiber flax (*Linum usitatissimum*)
tung (*Aleuritis fordii*)

Nuts
almond (*Prunus amygdalus*)*
pecan (*Carya illineensis*)

Vegetables
radish (*Raphanus sativus*)*
chard (*Beta vulgaris cicla*)*
chayote (*Sechium edulis*)
borage (*Borago officinalis*)*
burnet (*Sanguisorba minor*)*
eggplant (*Solanum melongena*)*
spinach (*Spinacia oleracea*)*
endive (*Cichorium envidia*)
asparagus (*Asparagus officinalis*)*
pumpkin (*Cucurbita pepo*)
leek (*Allium porrium*)
celeriac (*Apio napiforme*)
okra (*Hibiscus esculentus*)
cucumber (*Cucumis sativus*)

Condiments and Flavorings
black pepper (*Piper nigrum*)
anise (*Pimpinella anisum*)*
sage (*Salvia officinalis*)*
savory (*Satureia hortensis*)*
thyme (*Thymus vulgaris*)*
spearmint (*Mentha spicata*)
nutmeg (*Myristica fragans*)
yerba mate (*Ilex paraguatiensis*)
cinnamon (*Cinnamonum ceylanicum*)

Present Trends

The unmistakable trend today is for antique varieties of Old World crops and of certain New World domesticates to be eliminated from the agricultural inventory of the valley. Presumed absent are the root crops *maca* and *mauka* (LEON 1964a; REA & LEON, 1965); jackbean (*Canavalia plagiosperma*); yam bean and vanilla. Although one cannot prove that these crops were ever cultivated in the valley, extrapolation of crop patterns for the whole Central Andean region suggests that they were present in the Inca and colonial periods. Other cultigens, such as *llacón* and *achira* are now very rare.

This gradual but inevitable disappearance of the less efficient food producers is simply part of the world-wide acculturation of native peoples toward more modern, industrial frames of references. Greater ease of communications and the consequent frequent interchange of seeds, aided by the governmental agencies' distribution of seed, will rapidly change the agriculture that still prevails. The use of more fertilizers, introduction of modern machines, increased diffusion of the ideas of agricultural efficiency and advantages of monoculture in commerce all will tend to homogenize the agriculture. In addition the poorer, more isolated land in the valley is slowly abandoned, thus causing the demise of farms that perpetuated the old ways. The slow, laborious folk selection of crops is being replaced by experimental studies in controlled plots and in shiny laboratories. Diversification that took Indian farmers many hundreds of years to achieve can now be produced in several botanical generations.

The assault of various agencies and organizations and the pressures they exert for agricultural change are considerable, if unmeasurable. Among the official agencies has been the Ministry of Agriculture of Peru which has established experimental plots in the valley to test new varieties of crops and has made recommandations to large and small farmers. Another agent of agricultural change in the valley has been the National Institute of Agrarian Reform. A regional agency has been the *Servicio de Agricultura de la Cooperación de Reconstrucción y Fomento del Cuzco* ('CRYF') which granted credit to small farmers in the valley to buy fertilizers; the *Banco de Fomento Agrícola* also functioned for credit distribution. Since 1935 when several Italian agronomists were invited to study the Peruvian grain situation, representatives of foreign governments have made recommendations to modernize agricultural production and distribution in Cuzco. In particular, the United States, through the Agency for International Development and the Peace Corps, has supplied funds and manpower to introduce concepts of North American agriculture to Highland Peru. Many studies of the economics of agriculture in Peru, such as that of the Food and Agricultural Organization of the United Nations (Agricultural Development of Peru, 1959) have recommended the industrialization of indigenous crop and livestock production. Some of these reports, exemplified by COUTU & KING's The Agricultural Development of Peru convey a sense of urgency in the need to 'energize' Peru's agricultural sector, particularly in the Sierra.

Private groups have also instigated change of notable consequence in the valley. The Inca Schools Society, a British evangelical mission group, came to the valley early in the twentieth century. They have introduced steel plows, cultivators, harvesters, threshers, new races of cattle and sheep and hybrid seed from their centers in Calca, Sicuani and Urcos. The Cuzco brewery (Cervecería del Sur) has achieved important changes by contracting peasant farmers to grow two-rowed barley which is then sold to the brewery for making beer. As a means of eliminating the old six-rowed barley, considered to be inferior for beer, the brewery distributes small quantities of seed free to the peasants. Another private organization, the Internado Salesiano, a trade school run by Italian priests near Yucay, strives to teach peasant children European techniques in agriculture.

Success in altering the native agriculture is not necessarily a function of sociocultural level. Many Indians in what would be considered (for the region) relatively prosperous communities, have adapted many European ideas in agriculture, both plants and methods. Instead, acculturation of agriculture reflects economic levels; those poor peasants, Indians or cholos, living at the very margins of livelihood are the most conservative in their agriculture. In many cases, they cannot afford to make innovations because seeds must be purchased with money, because there may be no market for a new cultivar or new crop they may attempt to grow and because they are generally unfamiliar with the risks entailed in growing something different. With the growth of cooperatives in La Convención, farmers there are accepting crop and crop varietal change more rapidly than those living in other parts of the valley.

NON-DOMESTICATES ASSOCIATED WITH MAN

Non-domesticates associated with man include cultivated undomesticated species, unintentionally cultivated plants (weeds), and uncultivated camp followers (ruderals). Plants in each category may be either native to the New World or introduced from the Old World.

Cultivated non-Domesticates.

Cultivated non-domesticates are cared for in much the same manner as are crops, that is, they are weeded and protected from animals, often irrigated and sometimes harvested. They may or may not be planted by man. Those that are planted by man include, in the valley above 2,600 m, many of the larger trees which are valued as sources of lumber, fuel or as ornamentals. It is possible to infer that some species were transferred from the wild to the cultivated status in pre-Conquest times. Pisonay (*Erythrina falcata*) is a tree native only to the ceja de la montaña, but it is found growing as an ornamental shade tree in many town as high as 3,400 m. Cedro (*Cedrela Herrerae*), also native to the ceja between 2,300 and 1,800 m, experienced a similar ecological transfer and is now cultivated between 3,400 and 2,800 m in plazas and gardens. More utilitarian is kishwar (*Buddleia incana*) which is planted near huts and chacras up to 4,000 m to provide wood for agricultural tools. Now very rare in the wild, the native ecological position of kishwar was the temperate valley. The South American willow, *Salix Humboldtiana*, has been planted along stream sides since Inca times to furnish osier for making baskets. Old World species are also in this category; *Eucalyptus globulus*, native to southeastern Australia and Tasmania, was essentially unchanged by its transfer from Australia but in America it functions as a crop.

Other cultivated non-domesticates may actually be relicts from natural vegetation, now long cleared, which were not destroyed because man valued them enough to preserve them. Although they are not planted, they clearly have a protected status. Witok (*Genipa americana*) with a valuable fruit, and algarrobo (*Prosopis*) with pods used as fodder are examples.

Adventives

Adventives (ANDERSON, 1939:364-5) include both those plants unintentionally grown by man in his fields and gardens (weeds), as well as those pioneers that spread into habitats created by humans or by natural geomorphic occurrences away from crop fields (ruderals). Weeds are often inadvertently reaped with the harvested crop and the seeds are thus mixed. The peasants are careless about cleaning the seed, so the weeds are sown along with the crops. Ruderals spread into such man-created habitats as barnyards, roadsides and badly eroded places, or may come into an area after a landslide or shift in a river channel. From the point of view of man's attitude and use, both weeds and ruderals may be arranged in a continuum from protected through tolerated to rejected (Fig. 27). Most of the important weeds and ruderals in the valley are native to the Old World and were introduced into the valley in the sixteenth century about the same time as the crops. Some, however, came later and others continue to arrive. The ecological similarities between the Sierra and the Iberian Meseta assured the success of European adventives in the valley. Domesticated animals undoubtedly aided the diffusion of many adventives, creating new open habitats by trampling with hooves, dispersing fruits and seeds caught on their hair and depositing seeds in their droppings.

Because of their ubiquity, few adventives, although valued, are given a protected status. Among the ruderals given this special consideration are *Nicotiana paniculata*, a ruderal tobacco seen in barnyards; *Spartium junceum*, an Old World shrub used for thatch and in folk medicine; *Mentha arvensis*, a mint used in cooking; and *Matricaria chamomilla* used as a tea.

Many useful weeds and ruderals are allowed to maintain themselves up to a point because they have some value to man. Some weeds, such as *Amaranthus quitensis, Chenopodium petiolare* and *Brassica campestris,* are highly valued as human food, particularly the very young leaves. *Brassica campestris* may eventually be domesticated if present practices continue as many depend so much on these weedy potherbs that they may intentionally select them. Domesticated animals, being less discriminating than man about food, eat many more adventives. Those useful as fodder include such weeds as *Trifolium amabile, Sonchus oleracea, Raphanus sativus, Poa pratensis* and *Convolvulus crenatifolius,* as well as the old and tough *Amaranth, Chenopodium* and *Brassica* weeds previously mentioned. A few weeds, such as *Argemone mexicana*, are valued in folk medicine, but more ruderals than weeds have such uses. *Hydrocotyle alchemilloides* is a tolerated ruderal (but sometimes also becomes a weed). *Spartium junceum* is a tolerated ruderal in Urubamba and Calca Provinces because it is a source of thatch and medicine. Others are the shrub *Jatropha Curcas*, a source of firewood; *Fragaria vesca*, wild strawberry, a source of food; *Ricinus communis*, castor bean, used for firewood and the shrubs of *Baccharis* for firewood.

In a folk society such as that of the Vilcanota Valley most elements of the adventive flora have some compensating value, and few plants are totally objectionable. Yet man rejects some adventives either because their usefulness is minimal or their characteristics that make them pests outweigh their redeeming qualities. In some areas of the valley, kikuyo grass, *Pennisetum clandestinum,* has so aggressively taken over fields that any value this plant has for animal fodder is far outweighed by the land it takes away from crops. A few adventives are not only so rejected but even so noxious to the Vilcanota peasant that when encountered, they are destroyed wherever they may be. Peasants fear plants that kill livestock, such as *Nicotiana glauca* and *Astragalus garboncillo*; also considered undesirable are certain parasitic plants that live on crops such as dodder (*Cuscuta grandiflora*) and mallanway (*Phrygilantus ellipticus*).

WILD FLORA

The climax and subclimax wild flora differs from plants in the above categories in that it is primarily native in origin. Wild plants, together with wild animals, furnished the only means of subsistence for man in the Vilcanota Valley before the arrival of incipient agriculture. But today the wild flora plays a smaller role in human life in the valley than do crops and those non-domesticates associated with man. The wild native plants that had particular value and could gradually adapt themselves to disturbed habitats are now associated with man and his works, and are therefore classed under native crops, weeds and ruderals. Through the centuries man has changed the nature of the wild flora by burning and cutting, but he continues to depend on it for construction materials, folk medicinal remedies and forage. In no case, however, does man now depend solely on the wild flora for his subsistence although he may seek wild sources of food on occasion.

Plant Use in Peasant Life

The variety of ways in which man uses the array of plants in his environment depends on his cultural heritage, technological level and economic orientation. The peasants in the Vilcanota Valley look to the plant world not only for food for themselves and their animals, but also for shelter, heat, light and medicine. They also use plants to make many tools, for ornamentation and to indicate property boundaries.

Since the Conquest, Old World plants have increased the total inventory of useful plants; nevertheless, man in the valley respects tradition more than innovation and he continues to use most of the plants he had before the Conquest. The proportion of native to introduced plants varies from place to place in the valley depending on tradition, preference and a lack of alternatives. As seen in Table I below, Inca patterns of plant use are most faithfully followed in the puna environment and in the tropical and subtropical section of the valley. In the temperate mid-section more Old World introductions are used, but even here man depends most heavily on New World species, not all of which are native to the Central Andes nor necessarily present in the valley in the pre-Conquest period.

PLANTS USED AS FOOD

At least 90 percent of the food eaten by Vilcanota Valley peasants in normal years is supplied by cultivated plants. The remainder comes from animal sources and uncultivated plants. Staple foods are those consumed year round. Seasonal foods are important additions to the diet only in certain months of a year. Festival foods are eaten for special occasions as after the harvest. Famine foods are eaten only when drought, flood or disease destroy the normal crops. In addition to the grain and tuber staples, the valley peasants consume some vegetables, fruits, condiments and beverages made from plant products. Both native and introduced cultivated plant foods are important, but the proportion varies from place to place in the valley. The people themselves usually grow the foods or obtain them by local trade rather than by importing then from distant areas. The specific foods are consumed primarily because the people like them. Although poor and undernourished, these folk will not eat just anything, and have certain prejudices; deeply-rooted food preferences help explain why some innovations are rejected. The stock response to why such and such a crop is not grown is that the food is harmful (*'se hace daño'*), which in reality is more an expression of food preference than of health.

Staple Foods

Seeds and root crops are the staple foods of the peasant population. For the peasants of the upper valley, maize ranks first, potatoes second and wheat and barley, third. In the lower valley, the prime staple is probably manioc, although maize is also important. Cultivated seeds include the cereals (mainly maize, wheat and barley), the pseudo-cereals (mainly quinoa) and legumes. The grain of the maize plant in the valley is always used as food for people rather than to feed livestock as in the United States. Today peasants regard wheat highly, mainly because they have acquired a taste for bread. Oats are insignificant and rye, sorghum and millet are virtually unknown. Rice has become a food of some wealthier people and almost all of it is brought from the Coast. The seeds of quinoa, kañiwa and amaranth are less and less eaten. Pulses (known in Peru as *menestras*) are now of considerable dietary significance above 3,000 m; the most important of these is the broad bean, a native of the Old World. Usually pulses are dried, but they may also be eaten fresh in the immature stage.

Another important group of plant foods in the valley are those crops which are grown for their starchy underground parts. In the upper valley the potato in its many different varieties is the most important of these. Considering the Central Andes as a whole, the potato is the outstanding source of food. Other edible tubers of the upper valley are añu, oca and ullucu; all three, however, are slowly being replaced by more efficient and better tasting food crops and many people in the valley today eat them only sparingly.

Below 2,400 m other root crops take precedence. In the rainy section of the valley, arracacha is the most important food while farther down the valley, below 1,600 m manioc reigns supreme with uncucha (*Xanthosoma*), dasheen or taro and sweet potato being less important. Achira and llacón are still known by most people in the valley but rarely grown and eaten. Some peasants in the lower valley buy potatoes brought down from higher elevations.

Vegetables

Most vegetables (*hortalizas*) present in the valley are those which the Spaniards brought to Peru from Southern Europe. Only the cucurbits and the tree tomato

Table 1. Origins of the Plants now used in the Vilcanota Valley: Estimated Percentages of New World Plants v.s. Old World Plants.

Category of plant use	Puna Environment (4314 m – 3800 m)	'Temperate' Area (3800 m – 2400 m)	Subtropical and Tropical (2400 m – 730 m)
Plant Food in General	90% NW – 10% OW	65% NW – 35% OW	65% NW – 35% OW
A. Roots and Tubers	100% NW – 0% OW	90% NW – 10% OW	95% NW – 5% OW
B. Grain and other Seeds	40% NW – 60% OW	65% NW – 35% OW	95% NW – 5% OW
C. Fruits	none	20% NW – 80% OW	15% NW – 85% OW
D. Vegetables (including adventive potherbs)	95% NW – 5% OW	30% NW – 70% OW	90% NW – 10% OW
Material used in Construction	100% NW – 0% OW	50% NW – 50% OW	100% NW – 0% OW
Domestic Utensils	100% NW – 0% OW	50% NW – 50% OW	100% NW – 0% OW
Agricultural Tools	100% NW – 0% OW	95% NW – 5% OW	90% NW – 10% OW
Fuel	95% NW – 5% OW	30% NW – 70% OW	95% NW – 5% OW
Native Medicine	90% NW – 10% OW	70% NW – 30% OW	80% NW – 20% OW
Hedgerows	none	95% NW – 5% OW	100% NW – 0% OW
Ornamentals	100% NW – 0% OW	80% NW – 20% OW	70% NW – 30% OW
Dye	100% NW – 0% OW	80% NW – 20% OW	70% NW – 30% OW
Forage	80% NW – 20% OW	70% NW – 30% OW	65% NW – 35% OW

(*Cyphomandra*) are native. Three Old World vegetables have become widely grown and accepted as food: cabbage, onion and carrot. Other vegetable foods, such as lettuce and cauliflower, are grown occasionally for mestizos. Still others, such as artichokes, chard, endive, broccoli and Brussels sprouts, are almost never seen.

Fruits

Of the food products conventionally considered in this category, those of greatest importance are the banana, sweet orange, mango, sweet lemon and papaya from the lower valley; the pear, peach and capulí cherry from the upper valley. None of these fruits named is native to the Andes and all but two are introductions from the Old World. The native subtropical fruits, pakay, lucma, guava and cherimoya, have been subsequently relegated to a lesser position in horticulture and diet. In the upper valley very few native fruits ever grew. Most fruits are seasonal foods consumed only in certain months of the year.

Condiments

The most important native condiment is capsicum pepper, either fresh or dried. Indians depend on the pungent fruits to season food more than do cholos, but nowhere in Peru do peppers flavor everything as they do in Mexico. The European condiments garlic and onion are now very widely used in the valley. Black pepper, common in North America, is virtually unknown. Leafy condiment plants, collectively known as *asnapa* in Quechua, are important seasonings for the peasant folk. Of the eight herbs commonly sold for this purpose in native markets, seven are of Old World origin: coriander, oregano, rosemary, sweet basil, common balm, parsley and mint. Absent are other Old World herbs such as tarragon, thyme, dill and sage. Today the important condiment herb of New World origin is wakatay (*Tagetes minuta*); less seen today are *paiko*, a chenopod and *muña*, a mint.

Beverages

The most popular fermented beverage in the valley is chicha made of maize, a mildly alcoholic drink common throughout the Central Andes (VÁZQUEZ, 1967). Rich in calories, chicha is sometimes made with barley in addition to, or instead of, maize. Other fermented beverages have been made from peanuts or manioc. A range of fruits from quinces to capulí cherries has been occasionally used for alcoholic drinks, some of which have also been distilled on a small scale (Memoria/Urubamba, 1874:385) but this practice seems to be dying out. Sugar cane juice is fermented into a drink known as *cañazo*; it is also distilled into a potent spirit, rum (*aguardiente*), that has become a necessary part of all village fiestas in the valley. The everyday social drink, however, is still maize chicha, although commercially bottled beer, made in Cuzco since 1909, threatens to replace chicha as the preferred drink of the cholo class.

Non-alcoholic beverages of plant origin include coffee (strong and highly sugared), tea, cocoa and herb infusions, the most popular of which is camomile. Yerba maté or Paraguayan tea is not used, although there is some indication that it was used before the introduction of coffee.

Sweeteners

Aside from wild honey, the Inca's only sweet was a stalk of maize. The main stalk (*wiro*) of growing maize was cut into long pieces and chewed for the sweetish liquid. Later the practice of chewing maize stalks was transferred to sugar cane which the Spaniards had brought. Crystalline refined sugar has usually been out of the reach of most peasants, but they have consumed crude brown sugar (*chancaca*) especially as a quick-energy food on long trips.

Uncultivated Plants as Food

Generally less than 4 percent of the food intake of the Vilcanota peasant is from uncultivated plants, which in ordinary times are not dietary staples. Above 2,400 m, the native flora is poor and does not contain many wild plants that are both edible and palatable. Below 2,400 m the flora is much richer, but relatively few species have any real value as human food. Except the mushroom which is categorically rejected as unfit to eat, there is little rejection of wild plants as food sources. Fruits, leaves, stems and roots are the plant parts consumed. Uncultivated potherbs, boiled and added to a stew, are used in largest quantities during October, November, December and January when food supplies from crops are very low. They supply calcium, carotin, niacine and vitamin C, and their contribution to an individual's health can be significant.

During times of famine when crops fail because of such natural disasters as drought, flood or insect plague, the peasants are forced to seek other foods to replace those which were lost. Various wild plants are then assiduously collected to supplement the meager crops. The countryside is scoured for anything edible and plants from the wild are used with extraordinary intensity. In the Inca period, storehouses were strategically placed so as to be accessible to any area in need of food (COOK, 1925:110). Although not verifiable, intensive wild food collection may have begun only after the Conquest when the food storage and distribution network of the Incas disintegrated. A lack of rainfall in 1693 (Analesdel Cuzco, 1901:184), for example, caused widespread famine in the zone and the poor ate all kinds of herbs. In 1840 TSCHUDI (1847:257) saw starving Indian children roam the fields eating the grass like cattle. The plants today used in the valley during such emergencies may be the same ones used then. Among the most important are leaves of watercress, chenopods (particularly *C. petiolare*), leaves and stalks of *Trifolium amabile*, and the aquatic plants – *Lobelia*, *Hypochoeris* and *Nostoc*. The underground parts which are eaten range from the bulbs of *Nothoscordum* to the rhizomes of the begonia. The young fronds of *Dryopteris* fern are edible and flowers of *Cassia* are said to be palatable. Today various public and private organizations provide some form of relief to the peasants in the event of famine, as occurred in the 1957 drought in Canchis. But even when food is contributed from the outside, distribution is often faulty and it is often more reliable for the peasants – who over the centuries have learned to make their own security – to search the hillsides for edible wild plants than to depend on outside sources which may never arrive.

Food Preferences

Food preferences tend to follow socio-economic lines in the valley. Some basic foods, such as maize, wheat and potato, are accepted by all classes whereas others, kañiwa, añu, and kiwicha, are identified exclusively with Indians. Tarwi and oca are eaten by both Indians and cholos, but llacón and achira are eaten by cholos and mestizos at special times only. Popcorn is generally not eaten by mestizos for class reasons. On the other hand, in Cuzco Department today all four classes eat quinoa and ullucu, although HERRERA (1941:4) mentioned ullucu as being 'Indian food' until the beginning of the last century. The broader acceptance of this root crop would indicate that food conceptions and preferences are not static. To some degree food preferences are conditioned by what is available in the environment. For example, everyone in the lower valley eats manioc but in the upper valley it is regarded as inferior food.

If one were to generalize about peasant food preferences for the whole valley, maize would rank first as being preferred by the greatest number, wheat second and potato third. Whether this is an actual taste preference or a matter of several hundred years tradition is hard to say. A clear taste preference for indigenous foods and methods of preparation are difficult to demonstrate without alternatives being available.

Nutrition

Carbohydrates are plentifully supplied by maize and starchy root crops, but protein intake is low and generally insufficient to maintain a satisfactory health level. Meat is not a regular item in the diet and alternate vegetable proteins – such as kidney beans – are not eaten in large enough quantities to compensate for the dearth of animal protein. Meat is eaten when a sheep, cow, ox, llama or alpaca is slaughtered because of old age or when one dies a natural death. In addition, fiestas are a traditional time for eating meat, particularly roast guinea pigs (*cuys*) which are kept just for these special occasions (GADE, 1967:223). Insects are more important as a food in this society than is generally recognized, but they provide no great protein source. Most the protein in the diet, therefore, must be supplied by grains – wheat, barley, quinoa and, to a lesser degree, maize – which contain only limited amounts of protein as compared to animal sources. Cheese and milk are consumed sparingly. Very little fat, either animal or vegetable, is eaten and probably never was for seed plants are not grown here for their oil as they are in Old World folk agriculture.

A variety of plants, including grains, supply vitamins and minerals. Fruit is a good source of vitamins, but less of it is eaten than one might expect judging from its abundance and low cost. In the lower valley, fruit has been thought to be a carrier of malaria, an idea (CHIRIBOGA, 1937:9) that reportedly began during the devastating epidemic of 1932 in La Convención. Cholos accept vegetables such as cabbage and carrots, but Indians do not consume them regularly. Unlikely sources of vitamins for both cholos and Indians are the green leaves of condiment herbs and the green stalks of maize, both of which are high in vitamin C. Dried seaweed brought up from the coast and sold in blocks is served in soups and helps to prevent goiter because of its iodine content.

Food Habits

The peasants in the Vilcanota Valley do not consume much solid food before about 11 a.m. On arising in the morning, they customarily take only an herb infusion. The main meal of the day is somewhat before noon, and may typically include corn mush, boiled corn, and boiled potatoes with lots of condiments. On special occasions, such as during sowing and harvesting, about 2 p.m. a dinner is brought into the fields where the men are working. This consists of roast guinea pig, toasted maize, broad beans, chile pepper and liberal quantities of maize chicha. In puna areas above 3,600 m maize is generally absent from the daily diet, and below 1,500 m manioc is often the source of starch. Chicha accompanies the meal and is also drunk in the afternoon as a refreshment. In the evening before retiring (about 7 p.m.), the regular fare is a stew and boiled maize again seasoned. Hunger pangs between these two meals are allayed by chewing coca.

An idea about food which still persists from colonial times is the 'hot' and 'cold' concept: a sick person suffering from a 'hot' ailment cannot take 'cold' foods, and vice-versa. Which foods belong to which category depend upon regional and sometimes even local traditions. Generally, however, milk, broad beans, corn on the cob, chicha and bananas are 'cold', whereas capsicum pepper, salt, lard, camomile and tree tomato are among those foods considered 'hot'.

Basic Dishes

The basic dishes are gruel, soup or stew. Gruel is commonly made by adding water and occasionally milk to a ground meal of seeds or of the few dried tubers used in this manner. If the meal is coarse the gruel made from it is *lawa*, if fine *api*; More savory than gruel is stew (*chupe*) which peasants favor. The basic ingredient may be either roots (e.g. manioc, arracacha or chuño) or seeds (tarwi, broad beans, peas, wheat). Capsicum pepper is almost always used and very frequently potatoes and *Brassica* leaves are added when in season. Soup is a thin broth with chunks of food cooked in it, and is very popular; most peasants have it several times a week. It is also probably the most healthful, for boiling destroys many of the bacteria which cause severe intestinal disturbances. The basic ingredients of soup vary: dehydrated potatoes, triturated barley or wheat, boiled and dried maize, potatoes, arracacha, meat and cucurbits. Ullucu tubers, cut into thin strips, and *Brassica* or cabbage leaves are added to the pot whenever available. Condiments are essential, particularly dried capsicum pepper, but also oregano, coriander, wakatay and rosemary. *Puspo* or salad is another common dish, consisting of cold green peas or broad beans, greens and condiments.

Food Preservation

Preserving tubers by dehydration is one of the great achievements of the aboriginal peoples of the Andean Highlands of Southern Peru and of Bolivia. This is accomplished by alternately exposing tubers to subfreezing temperatures and pressing them. The process yields a distinctive food, since the flavor and texture of the dehydrated tuber are not at all the same as those of the fresh tuber. Primarily the potato (*chuño*), oca (*k'aya*) and ullucu (*lingli*) are frozen and dehydrated in the valley above 3,400 m. A much less common method of preservation is by boiling and dehydrating, but this is done to the potato and to maize. Grains (maize, wheat and barley), capsicum pepper, sometimes pome fruits and rarely bananas are preserved by drying them thoroughly in the hot sun. Some foods (manioc, citrus, onions, mango, papaya, arracacha, avocado) are not preserved at all and must be consumed soon after being harvested.

Food Preparation

Food is cooked by boiling, roasting, toasting, baking, popping and frying. Food was boiled (*waiko*) in Inca times and this still is the favored cooking method. Maize (either on the cob or as individual kernels), manioc, sweet potato (the starchy kind), potato, oca, uncucha, dasheen, arracacha and occasionally achira are boiled. Maize, wheat or barley may be boiled in water with wood ashes to loosen the outer covering; the wet grain is then hulled (but not crushed) on a hand mill. Another indigenous method of cooking is roasting. Most tubers and roots are on occasion roasted in small earth-clod ovens or in the embers of the hearth. Potato, manioc, dasheen, uncucha, achira, sweet potato, oca and añu are thus prepared. Another pre-Conquest technique is to dry seeds by heating them without scorching them. These parched foods, usually toasted in large pottery vessels, include wheat, maize, barley, amaranth, broad bean, kidney bean, peanut and quinoa, Parched seeds may be consumed as is or they may be ground on a saddlestone into a flour. Unparched seeds are milled into meal or flour by a water-driven water wheel. Dough has been made since Inca times but the idea of baking the dough may have been brought to the valley by the Europeans. Today most peasants in the valley accept wheat bread or wheat and barley bread as part of their diet, although they may not eat it every day.

The Incas knew about popping corn, but because only a special (and uncommon) variety of maize can be popped, it is rarely done in the valley today. Amaranth, a relict crop, can also be popped. Foods were not fried in pre-Conquest Peru, and this has proved to be the great Spanish contribution to cooking technique in the Andes. Today many dishes are prepared in this manner, particularly by the cholo and mestizo classes. Frying is especially common in the towns where bottled vegetable oil or lard are available in most small stores. The rapidity with which foods cook when fried perhaps accounts for the widespread acceptance of this method in such a fuel-poor region.

PLANTS USED IN CONSTRUCTION AND MANUFACTURES

Plants are an important source of raw materials for dwellings, kitchen appurtenances, agricultural tools and other necessities of life used by the Vilcanota peasants.

The typical peasant dwelling above 2,500 m is a small adobe hut with no windows, a ridged roof and a low and narrow door (Fig. 28). If it is of

Fig. 28. Looking over a high adobe wall into the barnyard where an Indian is cutting up eucalyptus wood. Her windowless, tiled-roofed dwelling is at the left.

more than a single room, one room usually serves as a storage area for agricultural products. The other room is the sleeping, dining, cooking and living area. Furniture is generally lacking; some cholo dwellings contain a table, chairs and a bed, but Indians still eat from the floor and sleep on sheepskins and homespun blankets laid on the ground. A corral (*canchón*), formed by a stone fence or mud wall, is located in back of or to one side of the hut where livestock is kept. A 2-meter high platform (*pirwa*), made of cut logs or simply small tree trunks, is used to store fodder out of the reach of animals (Fig.29). Although occasional huts are made of

Fig. 29. A fodder-storage platform (*pirwa*) inside a courtyard at Ollantaytambo, 2,790 m. Remains of an Inca wall at the right.

stones (sometimes removed from nearby Inca ruins), most are made of adobe bricks. To make adobes, dry ground, moistened to the consistency of soft mud, and grass (usually *Stipa ichu*) are mixed by Indians treading it with bare feet for several hours. When it becomes the right consistency, the mud is scooped into a wooden casing mold which is smoothed, allowed to dry somewhat, and then cut. The still-moist blocks are then removed and placed in the sun to dry thoroughly. Adobes are made during the dry season when the air is dry and sunshine is uninterrupted. After the adobes are laid in place and the building is finished, the walls may be whitewashed. Whitewash is made of lime mixed with the liquid from the heart of the giant cactus (*Trichocereus cuzcoensis*) which serves as a gluing agent.

The roof of the hut is supported by the ridge pole, a large thick beam to which the smaller poles that serve as rafters are tied with grass fiber rope. Large beams are of eucalyptus or ocasionally of kishwar (*Buddleia*) or alder (*Alnus*). The rafters are usually the hard stems of *Phragmites* which grows in wet places and is sometimes cultivated. Roofing materials may be either thatch or tiles, and ocasionally sheet metal (*calamina*). Until the beginning of this century,the valley peasants used thatch universally but now tiles cover many of the village dwellings. For example, RAIMONDI (1898:247) in the mid-nineteenth century noted that the dwellings at Pisac were covered with thatch; today about three-quarters of them are tiled. Thatch gradually becomes waterlogged and must be replaced by new grass on the average of every 3 years. Still, Indians sometimes prefer thatch to tile roofs even when they can afford the latter, because they believe that grass protects the hut from lightning.

Fig. 30. Indians constructing a dwelling on the puna at 4,200 m. The charred *Puya* stalks at the left, used as beams, are the only wood available in this treeless area.

Thatch is commonly the ubiquitous *Stipa ichu*, but other species are also used: *Stipa pungens, Festuca dolicophylla, Cortaderia quila*, and the branches of *Spartium junceum*

In the less temperate part of the valley, other adjustments are made in house building. Above 3,900 m dried *Puya* stalks support the walls of the circular thatched huts maintained by shepherds (Fig. 30). Below 2,400 m, the materials and form of peasant dwellings differ somewhat from those above this elevation. Partial adobe construction is common – reflecting the highland affiliations of most of the people in the lower valley – but structures are sometimes entirely of wooden poles. Easy availability of wood in part explains this choice, but it is said that on many estates peasant renters were not allowed to build more substantial dwellings. Supporting beams are made from a large number of tropical timber trees; a favored species is palo santo (*Triplaris peruviana*). Another desirable building wood below 1,500 m is algarrobo (*Prosopis*), a river bank tree used extensively for hacienda structures. In the ceja de la montaña split logs of torok (*Cecropia*) are used as siding for many huts, primarily because it is so easy to cut. In the lower valley, the poles are tied together with pliable vines (*bejucos*) such as *Calonyction aculeatum*, instead of with grass rope which deteriorates quickly in the warm, humid climate. The roof is usually more peaked than at higher elevations, and while grasses (usually *Festuca* spp.) are used as thatch, so are banana leaves (Fig. 31).

Fig. 31. Typical peasant dwelling of the lower valley, taken around 1,010 m. Except for the peaked roof, it is essentially similar to the huts in the upper valley.

Household objects are commonly fashioned of wood. Agricultural products are stored in wooden bins (*marka*). The *wishlla*, an all-purpose ladle used in cooking, is made of chachacomo (*Escallonia*) or unka (*Eugenia*) wood. Clothes hangers (*warkana*) are made of kishwar wood. Almost all spindles and coffins are now made of eucalyptus wood. In the lower valley, mahogany and cedro are used in cabinet making, a post-Conquest use. Wicker baskets are made of willow, kurkur (*Chusquea* bamboo) and pispita (*Acalypha*). In the past Huayllabamba was the center of basket making in the valley (PAZ SOLDÁN, 1862:377). Leakproof containers are made from dried *Lagenaria* gourds and, now only occasionally, from the hollowed-out fruits of the pamuco (*Crescentia cujete*) tree. Metal bowls and plates purchased in the larger markets are replacing the folk-manufactured containers. Most vessels and containers used by the peasants, however, are crudely made earthenware.

Wood is the raw material for most agricultural tools. The woods of the kishwar and chachacomo trees are highly valued for the chakitaklla (foot plow), and the lampa handles and also pitchforks. Kishwar trees have been planted near huts just for this purpose probably since the Inca period. Also valued are woods of the waranway and lambrán trees for plows and yokes; tara for plow shafts; and old trunks of the cactus pata kiska for lampa hafts. Clod crushers are made of the flexible wood of the lloke tree.

Grasses and other plant fibers are twisted or braided into rope and twine. Some Indians in the puna make ichu grass twine and sell it in the Sicuani market to buy maize. Rope from llama fleece is also made above 3,900 m. In the past *Chusquea*, the bamboo of the ceja, was made into rope. Heavier rope is made in the valley from the fibers of two agave species: *Furcraea andina* and *Agave americana*. However, each year fewer and fewer peasants do this work. In the lower valley, certain unidentified vines known under the general name of *bejuco* serve as rope. The pliable branches of certain species in the upper valley, willow, *lloke*, agave and *Furcraea*, have been used to make suspension bridges. Today very few of these handwoven bridges remain in the Southern Sierra and none in the Vilcanota Valley, having been replaced by steel cables (GADE, 1972a).

Woven textiles in the valley are overwhelmingly made from animal fleece. The sheep, alpaca and llama supply most of the yarn for textiles because woolen clothing is more practical in the climate of most of the valley. Only in the lower valley is cotton used, but even here, it is spun only on a very small scale. The fibers of the *Ceiba* tree may once have been used to spin thread. Many cholas no longer weave or spin, so all their clothing (and footwear) are factory-made and purchased in markets. Indians, however, still weave their own textiles. To wash their homespun clothes, the bark of *rokke (Colletia)* and *Sapindus* have been used as detergents.

PLANTS USED IN FOLK MEDICINE

The Vilcanota peasants are heirs to a rich herbology which probably dates from before the Incas. Since the Conquest species from the Old World have been added to the inventory of medicinal plants, but the great majority of herbs still used are those native to the Central Andes area. Very few are purposely cultivated, some are adventives, but most are members of the native wild flora. Apparently the majority of plant species in the valley are thought to have some medicinal value. Some of these plants are of genuine therapeutic value and are listed in official pharmacoepias (including the U.S.P.), but others are not. Whether based on science or magic, certain species have gained the reputation in native pharmacology of being particularly efficacious. The valley below 2,400 m yields a great number of these, if only because the flora there is much richer; proportionately to the total flora, however, the forested areas do not contain more species regarded as particularly useful than the valley above 2,400 m.

Of some plants, only one part is valued; of others, several or even all vegetative parts – root, leaf, stem, bark, flower, seed, and fruit – may be useful, but not necessarily for the same ailments. The plant or a part of it may be used fresh or may be used only in a dried and powdered form. Fresh plants applied externally to treat wounds, bruises and sprains, or to alleviate nerve-caused pain, are made into poultices. Highly regarded as such are the leaves of *Nicotiana paniculata* (k'amasayri); *Oenothera rosea* (yawarchunka); *Senecio pseudotites* (maycha); *Grindelia boliviana* (chiri chiri); *Loasa ranuncilifolia* (orko kisa); *Eucalyptus globulus*; and *Hydrocotyle alchemilloides* (matek'lo). Also applied externally to treat wounds are vulneraries such as *Weinmania bifida* (wichullu); *Bocconia frutescens* (yanali); and *Piper angustifolium* (moko moko).

Most plants taken internally are usually in infusions. Sometimes a fresh part of the plant is used, but more often it is dried. Outstanding as diuretics are *Smilax officinalis* (zarzaparilla); *Krameria trianda* (pacha lloke); *Plantago major* and *P. hirtella* (both llantén); *Melissa officinalis* (toronjil); *Cybistax* spp. (achiwa achiwa); *Calceolaria cuneiformis* (puru puru); and *Ephedra americana* (pinco pinco). *Rumex cuneiformis* (llake) is a recognized depurative. *Eupatorium sternbergianum* (chamanway) and flax seed (*Linum*) are highly regarded for stomach disorders. *Cinchona* bark is highly regarded as a febrifuge.

Purgatives are important in this society in which almost everybody suffers from intestinal parasites. The following plants are used to expel worms: *Euphorbia penicillata* (wanchanka); *Jatropha Curcas* (piñon); *Solanum pseudolycioides* (t'ankar); *Psoralea pubescens* (wallwa); *Argemone mexicana* (k'arwinchu); *Sicyos Baderoa* (putak llanko); and *Aniba* spp. (pucherí).

Several plants are taken in infusion as remedies for non-specific ailments, such as *Matricaria chamomilla* (manzanilla); *Lippia Fiebrigii* (cedronsillo); *Foeniculum vulgare* (hinojo); *Mentha arvensis* (yerba buena); and *Minthostachys tomentosa* (muña). Several other plants have more definitely established purposes: *Ruta* to cure menstrual pains; *Micania* to cure snake bite; *Sambucus* for sore throat; *Xanthium* cures a toothache.

Certain plants are not intended to cure diseases or ailments, or to heal wounds, but are for certain other effects. Many plants in the valley are thought to have narcotic effects, but verifying that some generally recognized hallucinogens are used is extremely difficult. Wilka (*Piptadenia*) seeds were used in Inca times and may still be taken. The seeds of floripondio, *Datura arborea*, and k'arwinchu, *Argemone mexicana* are reportedly strongly hallucinatory but inquiries about their use met with stout denials. Hallucinatory mushrooms apparently do not grow here. Coca leaves serve as the outstanding narcotic in the valley and probably have since before the Inca period. Many people in the valley depend on coca to allay hunger and cold, and many peasants in the lower valley also depend on it for a cash crop. To extract the alkaloid from coca, calcium balls (*llipta*) made from the ashes of burned quinoa, kañiwa, amaranth, *Baccharis* stalks or cacao pods are chewed with the coca. In the lower valley, *chamire*, the root of an unidentified selva plant, is occasionally taken with llipta to sweeten it. At least one plant is used to stupefy fish: kumu, *Tephrosia toxicaria* Wanarpo (*Jatropha ciliata*) is recognized by many people in the valley as an extremely effective aphrodisiac.

The most common medicinal plants are rarely seen in markets, for the peasants themselves collect them as needed. For those that are rare, hard to find or products of distant localities, certain individuals make a living by collecting and/or selling them at special herb stalls. These herb stalls are located on the fringe of the main markets in the larger towns. Known in Quechua as *hampi katu*, they are veritable museums of Inca medicinal plants (together with rocks, powders and dried animals) and Spanish colonial additions. Good luck charms, such as wayruro seeds, are also sold. Cuzco has the biggest herb market in the Department. The biggest day of the year in this herb markèt is the morning of Good Friday when the medicinals sold are considered to have extra efficacy because of the religious significance of the day. Other regular herb markets in the valley are at Sicuani and Quillabamba. Ambulant Bolivian herb sellers (*kallisayas*) occasionally come to the smaller towns to sell their wares on the big market day.

PLANTS USED AS FUEL

In spite of its relative scarcity, wood remains the single most important source of fuel in the valley. In the larger towns, kerosene-burning Primus stoves are now popular and firewood vendors are no longer common in markets as they were until the 1940's. In the countryside, however, most of the peasants still cook on the hearth and for this they burn wood that they purchase or collect themselves. Charcoal making, apparently introduced to the valley by Spaniards, is no longer an important activity because the people who can afford to buy charcoal now have kerosene stoves. Occasionally cholos in the lower valley make charcoal, but Indians do not make it, at least not for their own use.

In the puna above 3,900 m trees are almost completely absent and vegetal fuel comes from combustible shrubs and grasses. Yareta, a resinous cushion plant of the umbelliferous genus *Azorella*, is the most valued fuel at these high elevations but it is now very rare in the Vilcanota Valley in accessible places. The most frequently used woody fuel is the dried inflorescenses, leaves and woody stems of *Puya*. Because good vegetal fuel is scarce in this section, dried llama, alpaca or sheep dung (*takia*) or cow dung (*k'awa*) is widely used. Dung burns quickly and gives a hot fire, an asset at these high elevations where oxygen is low and fires hard to start.

In the temperate section of the valley, from about 3,800 m to 2,400 m, eucalyptus plantations furnish much of the fuel used. Approximately 70 percent of the firewood in this part of the valley comes from *E. globulus*, which yields a good-burning wood leaving solid embers. Eucalyptus firewood is expensive, costing from 8 soles per quintal (about 45 kgs.) in areas where it is common to 12 soles where it is not. Poorer peasants who cannot afford to buy firewood comb the countryside for any combustible material. After eucalyptus, the next most important firewood species is molle ((*Schinus molle*). Other native tree species are now quite uncommon, probably because of centuries of cutting, so that they contribute little to the total firewood inventory of the valley. Certain shrubby plants are depended upon as combustibles; among these are tayanka and chilka, two resinous species of the genus *Baccharis*; hetteka (*Ophryosporus origanoides*); and mutuy (*Cassia* spp.). Dried grass and leaves are also collected for kindling (*charamusca*), as fires are difficult to start at these altitudes. Ovens take special fuels. The leaves and branches of eucalyptus are burned in ovens to bake roof tiles and in bakeries to bake bread; even more popular for this latter use is rokke (*Colletia*) which is regarded by many bakers as the only suitable fuel for bread ovens. (Fig. 32).

Combustible material is abundant below 2,400 m because of the plethora of trees. From 2,400 m to 1,500 m firewood is plentiful, but often waterlogged and must be dried. From 1,500 m to 730 m, good firewood is available only in quite

Fig. 32. Indian in Pisac carrying a load of *rokke* (*Colletia*) branches, an important combustible used in bread-making ovens.

inaccessible places, but there is by no means a shortage as in the upper valley. The more desired fuel woods are wilka (*Piptadenia*); algarrobo (*Prosopis*); chamba (*Leucaena*); and chaman (*Dodonea*). Several weedy shrubs such as *Jatropha Curcas* and *Ricinus communis* are burned extensively because they grow in easily accessible places.

PLANTS USED AS FORAGE FOR DOMESTICATED ANIMALS

Forage for herbivorous domesticated animals may be planted fodder or natural vegetation. The llama and alpaca get their provender from wild grasses and mosses, and there is no evidence that the Incas planted any crop primilarily as forage (Fig. 16). The teeth of llamas and alpacas are said to remain in better condition if the animal is raised on natural forage than if fed the softer cultivated fodder. It was not until the Europeans introduced cattle, horses, burros and mules that planted pasture became necessary, for these Old World domesticates do not thrive on the dry, sparse natural vegetation of the Sierra.

The three major planted forage crops in the valley since the beginning of the colonial era have been alfalfa, green barley and maize leaves and stalks. The importance of pack animals in the period 1600-1900 must have required more

Fig. 33. Indian carrying weeds out of the maize field in January; these weeds will serve as fodder for the cow.

planted forage that one finds today. When the Cuzco-Potosí trade flourished, Tinta was the center of an important mule raising area. In the seventeenth century (JUAN Y SANTICILIA & ULLOA, 1752:512), 25-30,000 mules brought from Tucumán (in Argentina) were grazed on pasture land in the Corregimiento of Canas and Canchis (present-day Province of Canchis) before being sold at the yearly fair. Since mules do not thrive on the poor natural grass during the dry season, fodder must have been planted, and it is likely that the Vilcanota Valley had considerable land devoted to forage crops for these pack animals. Another rise in forage hectarage probably occurred after 1821 when the Spaniards withdrew and Indians were, for the first time, permitted to keep their own horses. Later, after motor trucks replaced animals as the primary means of transportation, one may assume that planted forage crops in the valley diminished. Today some alfalfa is planted to feed hacienda-owned cattle, but most peasants do not grow it for their own animals. Cholos and mestizos occasionally grow barley on a small scale for their favorite domestic animals. Wheat straw, barley straw, maize *chala* and stems of potatoes and other tuber crops are fed to domestic animals. The stubble of crops is grazed after the harvest; a practice which also fertilizes the fields by the animal droppings.

Weeds and ruderals are also used to feed livestock. A field may be weeded several times in a growing season and the carefully collected adventives are then given to cattle and other animals (Fig. 33). Among the weeds most used as forage plants are convolvulaceous bindweeds, leguminous weeds particularly *Trifolium amabile, Melilotus alba* and *M. indica*. Other adventives which are sources of fodder are wild oats (*Avena barbata*), wacha (*Rumex crispus*); shepherd's purse (*Capsella bursa-pastoris*); sow thistle (*Sonchus oleraceus*); and willko (*Mirabilis prostrata*).

Mostly the peasants leave their animals to scrounge for themselves, finding whatever they can to eat. In part this practice reflects that the crop is precious and needed by humans, but it reveals also the culturally conditioned attitude that livestock can satisfactorily subsist on natural hillside pastures and riverbank vegetation, even during the dry season. Although several wild grasses have forage value (species of *Stipa, Festuca, Paspalum, Poa* and *Bromus*), the scrawny undernourished appearance of most domesticated animals in the valley indicates the generally poor quality of the browse. A notable exception is between the villages of San Pedro and San Pablo in Canchis. Here natural rush meadows grow on poorly drained valley bottom land. Although a simple engineering readjustment could drain these meadows and make good crop land, the Indians do not want these lands dry for they supply ample forage for cattle at all times of the year. The cattle of this area are notably superior animals to those where the browse is not nearly so rich, even though most of them have liver disease transmitted by snails living in the meadows.

PLANTS USED AS BARRIERS AND BOUNDARY DEMARCATIONS

Certain plants are used in the valley to mark property boundaries and to serve as field barriers. Most of these so used are wild, favored by man who plants them and reduces or eliminates their natural competition. Many reproduce vegetatively thus simplifying their propagation.

In the Inca period there was no great need to define field boundaries since the land was communally held and its user changed often. With the Conquest and the subsequent establishment of private properties, no land surveys were made to determine property divisions and no deeds were drawn up to define and record them in an accurate, mensurable fashion. Boundary disputes were inevitable and the bitter litigation which followed usually favored, as it still does, the large land holder who speaks Spanish, has powerful friends and is familiar with the legal intricacies. Because the Indian possesses only small amounts of land, he is warranted in guarding them extremely closely for a change in a boundary line could dangerously reduce his food supply. Vegetal plantings mark *de facto* property limits and thus lessen the chance for boundary disputes. In the upper valley, particularly in Canchis, pakpa (*Agave americana*) marks the boundaries of many parcels. In addition, certain plants function mainly as fences to keep people and livestock out of crop fields. Thorny plants (*k'aru*), such as cacti (mainly *Opuntia exaltata, O. perrita* and *Trichocereus cuzcoensis*); llaulli (*Barnadesia horrida*); barberry (*Berberis boliviana*) and rokke (*Colletia spinosissima*) are often densely planted on top of a low adobe wall.

Hedgerows are small trees and shrubs densely planted in a linear pattern. They may define property boundaries, act as fences or provide materials needed for tools, baskets or construction. Hedgerows are, therefore, an efficient use of space. They are especially common between 2,800 and 2,500 m where the common species are *Sambucus peruviana, Schinus molle, Tecoma sambucifolia, Cassia* spp., *Caesalpinia spinosa*, and *Stenolobium stans*. In the lower valley below 1,500 m the natural second growth is often left between parcels and thus man-made hedgerows are less conspicuous in this landscape.

PLANTS USED AS ORNAMENTALS

Many flowering plants have ornamental usage, but a few are especially valued for adornment, either for festive occasions, death rituals or as decoration around the living area. Ornamental flowers from the Inca period are known from painted designs of plants on pottery and wooden cups (VARGAS, 1962). Motifs of kantu (*Cantua buxifolia*), ñukchu (*Salvia tubiflora*), and chinchirkuma (*Mutisia*) blossoms are especially apparent on these Inca artifacts. Clothing, particularly women's skirts and blankets, had a stylized fern design which is still used today by some tradition-bound Indian weavers.

Ornamentals may be wild, protected or cultivated. Native trees, shrubs and herbs may be planted for simple adornment; these include pisonay (*Erythrina falcata*) with its brilliant orange-red papilionaceous flowers, red-flowered *Fuchsia* and floripondio (*Datura arborea*) with its large white pendant blossoms. Also growing around huts is the dahlia and, in the lower valley, bougainvillaea. The Old World cypress is seen planted in some villages. Flowers may be cultivated as a cash crop or for use in certain festivals. The main commercial flower growing center in the valley is at Huaraypata in Quispicanchis where gladioli, madonna lilies, dahlias, cosmos, carnations and other flowers are grown under irrigation in peasant enclosures. Small clusters of tarwi (*Lupinus mutabilis*) are sometimes planted for decoration, and as its use for food declines, the future of tarwi may be as an ornamental (GADE, 1969:49).

Certain native plants, while not actually cultivated, may receive protection because of their attractive blossoms, and some are sanctioned by Quechua tradition as ornamentals. These include kantu (*Cantua buxifolia*), the national flower of both Peru and Bolivia; chinchirkuma (*Mutisia*); and ñukchu (*Salvia*); possibly pfallcha (*Gentiana primuloides*) also fits into this category. On the joyous occasions of sowing and harvesting, men and women wear flowers of these aforementioned species in their hats. Also a bouquet of flowers is often placed on top of the pile of wheat seed that is about to be sown. The yellow blossoms of *Cassia* and *Spartium* are essential to the celebration of Godparents' Day.

In the valley a special use of wild plants is to decorate creche scenes (*nacimientos*) at Christmas time. From hillsides Indians collect both non-flowering and flowering plants which they bring to sell in the Cuzco market on December 24. These include several kinds of bromeliaceous plants, several species of ferns, club mosses, lichens, cactus flowers, passion flower, begonia, retama, lupine, nasturtium, as well as leafy branches of the capulí cherry, kishwar and lambrán.

PLANTS USED AS DYESTUFFS

Several native and introduced plants in the valley contain pigments which were used as dyestuffs in Inca and colonial times. The Quechua language has an elaborate terminology for all aspects of the dyeing process and we may suppose that in past times it was a very important activity. In the 1860's, chemical dyes produced in Europe began to compete with natural dyes, and by the twentieth century many of the natural dyes were abandoned. Among the natural plant dyes used before the aniline onslaught were a bluish-black pigment from the fruit of *Genipa*; a black dye from the leaves of *Kageneckia*; yellow dyes from *Bocconia, Eupatorium* and *Bidens*; red or orange dyes from *Relbunium microphyllum* and *Buddleia longifolia*; and a blue pigment from the leaves of indigo. Cochineal, a scale insect living on *Opuntia ficus – indica*, was the most important source of red dye in Inca and colonial times but it seems to be absent from the valley today. The ashes of several plants have been used as dye mordants, mainly molle (*Schinus molle*) and chaman (*Dodonaea viscosa*).

Several natural plant dyestuffs remain in use in the valley. The brown pigment from walnut leaves is used to dye textiles. Certain yellowish dyes are used primarily to color food for festive occasions. Palillo is the name given to a yellow pigment obtained either from the root of turmeric, an Old World plant, or from the root of a native scrophulaceous plant that is collected from the wild. Annatto or achiote (*Bixa*) is a semi-cultivated shrub yielding a red, orange or yellow dye, depending on the concentration of pigment. In the upper valley, a reddish-purple dye for food and drink is still made from the seeds of ayrampu (*Opuntia soehrensii*).

PLANTS USED IN TANNING

The pods of tara (*Caesalpinia spinosa*) continue to be important as a source of material for tanning hides. The bark and wood of queñuar (*Polylepis incana*) is also well known for this use, but this tree is now rare in the valley.

Distributional Analysis and Landscape Synthesis of Useful Plants

The spatial dimensions of useful plants in the Vilcanota Valley may be approached on three different scales: 1. crop assemblages in individual fields or gardens; 2. crop-growing farm units; and 3. complexes of useful plants by ecological zones. At the largest scale is the micro-distribution of plants within a field or garden. On a somewhat smaller scale is the grouping of crop fields comprising single agricultural units, either peasant farms or manor estates. At a higher level of generalization are the broad ecological regions of the valley each of which is characterized by the cultivation of a different combination of useful plants.

1. CROP ASSEMBLAGES

The heterogeneity of species found within a single garden or field is one manifestation of indigenous polyculture that is so deeply ingrained in peasant custom. The commercial agriculture of the estate owners usually has relatively little botanical intermingling because these producers are committed to Western ideas of agriculture which include ease of sowing and harvest, row cultivation and close sorting of seed. In peasant agriculture, a much higher degree of complexity of species exists within a single garden or field, yet among folk farmers themselves there are differences. Where hoe agriculture is practised rather than plow agriculture, a greater diversity of plants grow together. In addition, as one moves toward the warmer parts of the valley, there is also a tendency towards greater complexity and also a layering of useful plants. A garden chacra at 1,950 m in the ceja de la montaña shows a wide assemblage of plants grown together in one small area

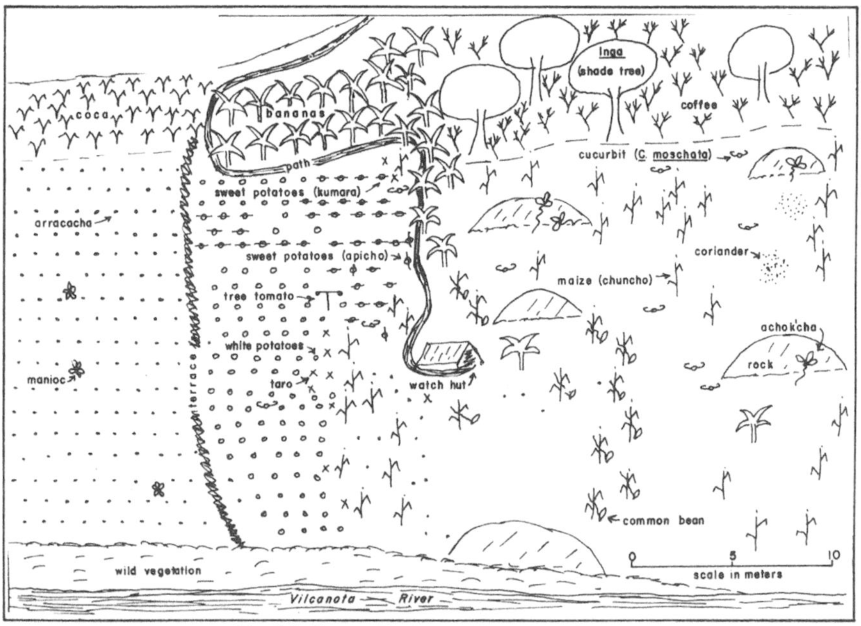

Fig. 34. Chacra Garden at Elevation 1950 m.

(Fig. 34): seeds and tubers, annuals and perennials, warm land crops and temperate land crops and plants having, beside use as food, beverage, narcotic and condiment uses. Although some concentration of one kind of crop into a particular part of the field is apparent, considerable interdigitation of plants occurs particularly in the middle of the chacra. Farther down the valley, around 1,000 m, a garden chacra would not contain potato and arracacha but would contain pineapple, achiote, papaya, mango, pakay, and Gros Michel (or seda) banana.

In the upper valley (above 2,500 m) are fewer species and less layering. The efficient use of space known in Mexico whereby beans use corn stalks for support and squash is the ground cover crop is not a common arrangement. In the Vilcanota Valley the layering and mixing is more hodgepodge than this: broad beans and quinoa are common in maize fields, potatoes are grown with other kinds of tubers and even in fields of small grains several cultivars of one species or a mixture of several species occur. One small 5 meter by 4 meter hillside field above the town of Yucay (2,960 m) yielded, in addition to a plethora of tolerated *Brassica* weeds, the wheat cultivars *mentana, florencia aurora* and *barba negra* as well as *común* barley.

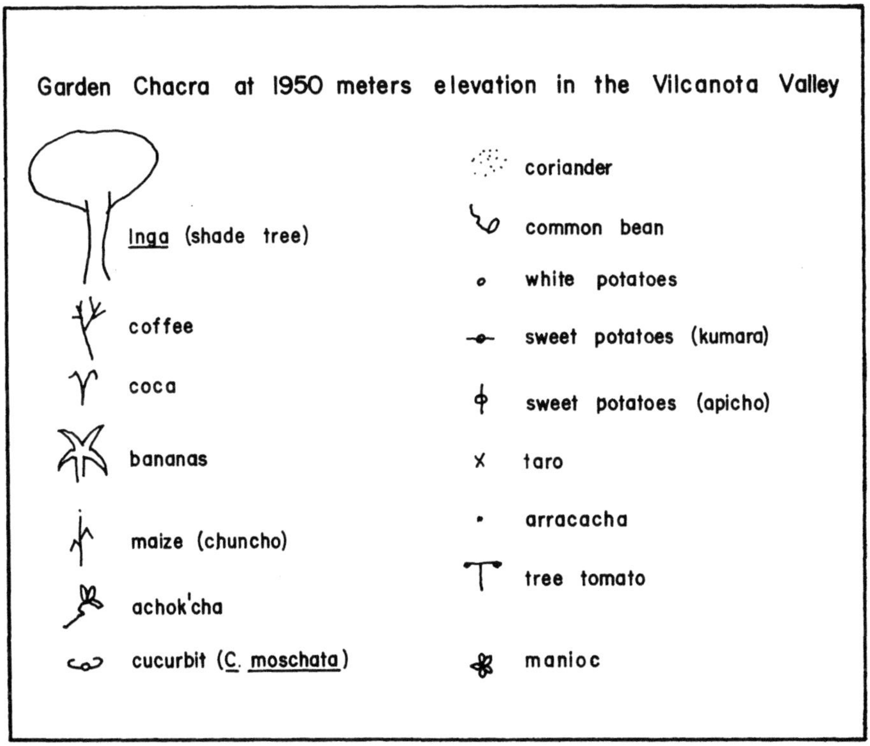

One frequently finds lower yielding, less efficient food producers, usually Andean in origin, planted on the periphery of a field or garden. Tarwi, achira, quinoa, añu, oca and ullucu formerly occupied whole fields in the valley before plant introductions pushed them to an inferior position; now they are often placed on the most exposed parts of the field, presumably because man now values them the least. Conversely, those crops which are most apt to be stolen because they are especially desirable and/or rare (such as *chullpi* maize), are planted in the middle of the field in an attempt to conceal the plant from prospective thieves (GADE, 1970c:11).

2. CROP-GROWING UNITS

The composite of fields reveals the relative importance of each crop within the

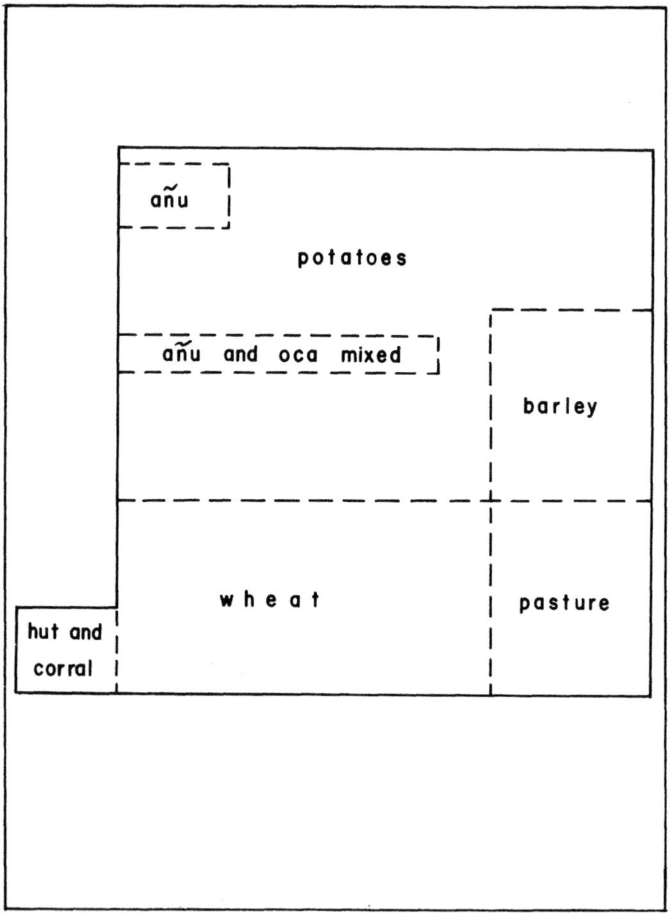

Fig. 35. Crop Arrangement in a Chacra near Chectuyoc, 3,600 m.

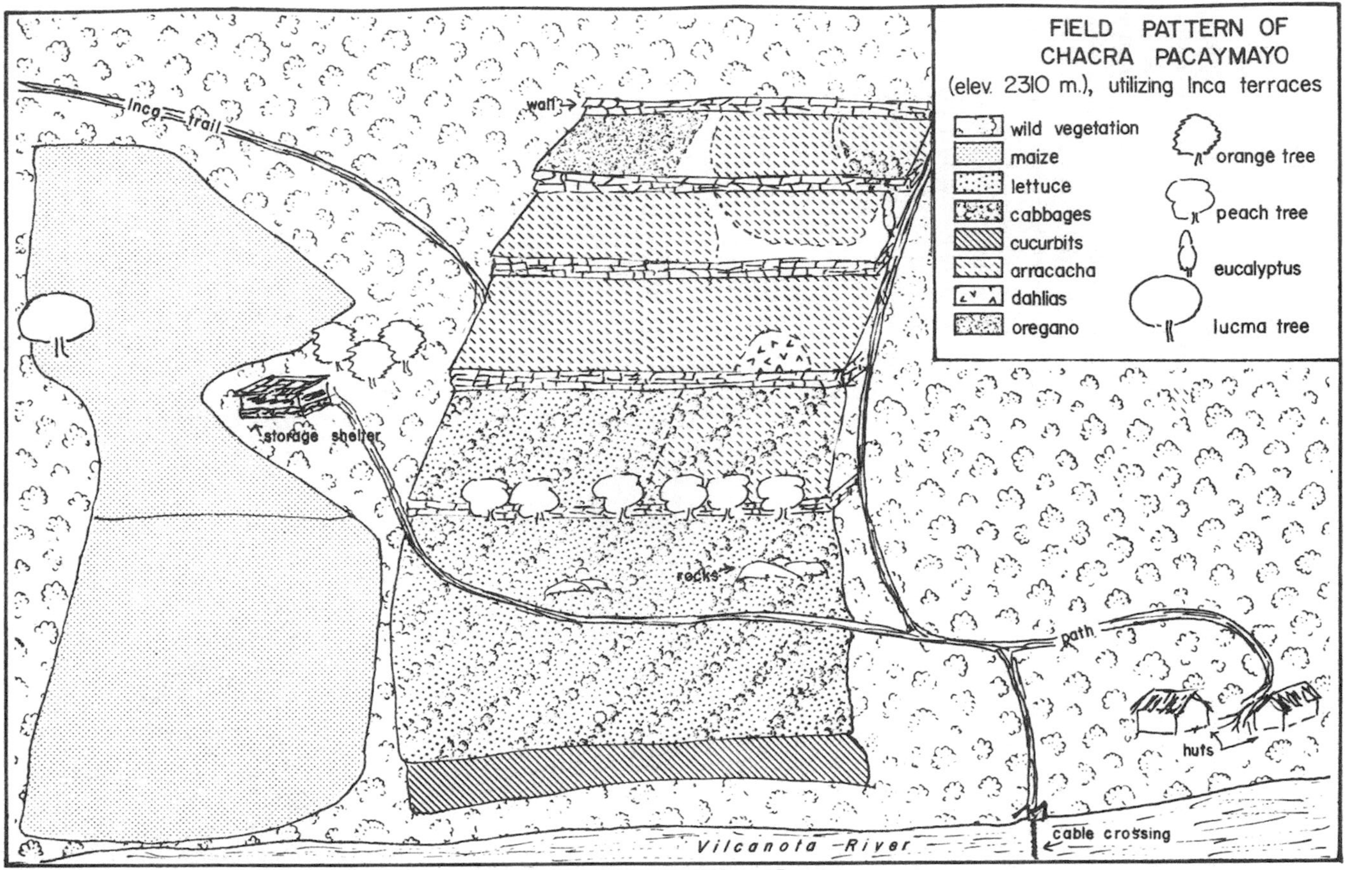

Fig. 36. Field Pattern of Chacra Pacamayo.

producing unit, whether it be subsistence or commercially oriented. Greater attention to one crop over others may reflect a cultural preference, an economic consideration, or a limitation imposed by the physical environment. The greatest array of crops is grown by peasants who make of this agronomic diversity a form of crop insurance. If one or even two of the crops should fail, several others will ordinarily succeed and thus the spector of famine will be kept away. Manor estate producers, on the other hand, can take more risks with monoculture because they do not operate at such a close economic margin. Some haciendas are so devoted to a single crop that they must purchase most of their food in the city of Cuzco. Three examples of crop patterns on areal units illustrate the above difference; the examples include two peasant farms 1,200 m apart in elevation and a fairly typical manor estate in the upper valley.

The first peasant farm, less than one hectare in size, has a cooler mesothermal climate and is in the valley bottom at an elevation of 3,600 m near Chectuyoc. The pattern on this farm is one of rectangular fields in which approximately 60 percent of the land is in tuber crops, 32 percent in grains and 8 percent in fallow pasture (Fig. 35). Most of the tubers are potatoes (*bole* and *kompis* cultivars) with a few intermixed rows of añu and, to a lesser degree, oca. The grains are wheat (*candaño* and *marquis* cultivars) and barley (*común*). The pasture area is actually a weed

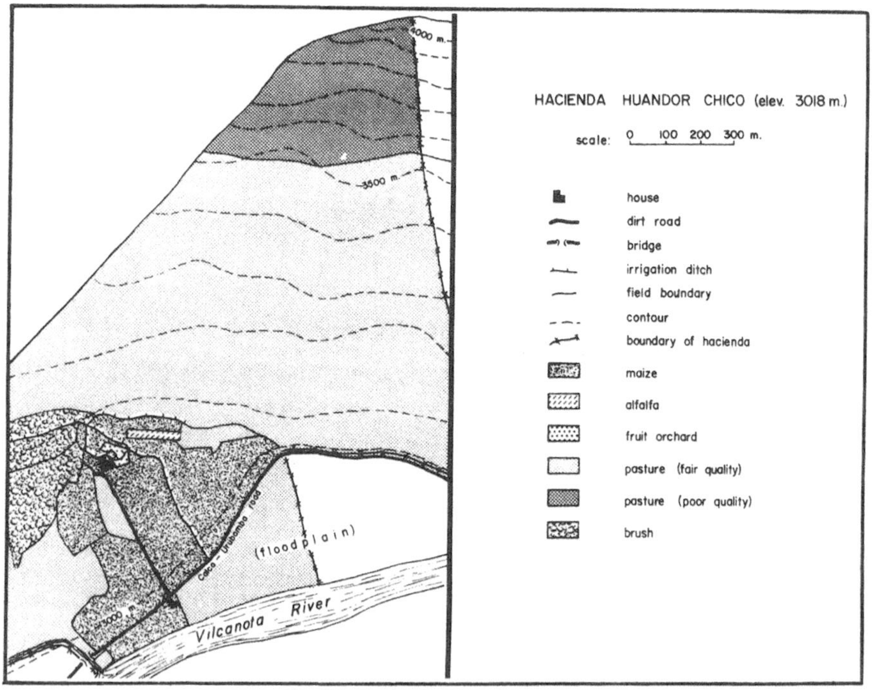

Fig. 37. Hacienda Huandor Chico.

patch in fallow to be plowed and sown to a crop the following year.

The second peasant farm was established only in 1962 by a cholo from the Cuzco Valley. This farm is at 2,350 m at the upper fringes of the ceja de la montaña; though drizzle is common enough here, the low-sun dry period still must be contended with and crops are grown unirrigated (*temporal*) only until the farmer can build irrigation ditches. Because the land was only recently cleared of forest, the land still remains naturally fertile and no fertilizers are applied. Of the $1\frac{2}{3}$ hectares in crops (Fig. 36), about $\frac{2}{3}$ hectare is planted in maize. One field of maize is of the upland flour cultivar, and the other is the semi-hard *laurel* maize of the ceja and lower valley. During any one harvest, one of the two types succeeds better than does the other, and some crop of maize is thus assured. Approximately $\frac{1}{3}$ hectare is in arracacha which grows here better than any other crop but which is not the family's preferred food. Another $\frac{1}{3}$ hectare is in other crops, mainly cabbage, lettuce and oregano which are sold when the market price is high in the lower valley. Squash (*Cucurbita moschata*) is grown near the river bank mainly to be fed to hogs but the peasants also eat it when immature.

The crop pattern of a manor estate is illustrated by Hacienda Huandor Chico, occupying the valley bottom and slope near Calca, at an elevation of 3,018 m (Fig. 37). On this estate only 24 hectares out of a total of 142 hectares are cultivated by the hacendado. Of these 24 irrigated hectares, 23 are in maize (*uwina* cultivar) and 1 is in alfalfa and fruit. His small herd of cattle is fed alfalfa as well as allowed to graze the permanent pasture of the slopes and the floodplain of the river. Overgrazing has led to the formation of brush land (*matorral*). The peasant families who live and work on the estate use the unirrigated slopes of the hacienda for small unirrigated plots and for grazing their animals. Above 3,500 m the pasture deteriorates into stiff, dry forage but the peasants are expected to use this pasture and leave the higher quality pasture below for the cattle of the hacendado. Manor estates elsewhere in the valley concentrate on a saleable crop or crops; above 3,500 m this tends to be the potato, while below 2,000 m it may be tea, coffee or sugar cane.

3. COMPLEXES OF USEFUL PLANTS BY ECOLOGICAL ZONES

The physical variables, temperature, moisture, hail, length of growing season, singly or in combination, partially control crop distribution. Essentially two crop limits exist: the effective limit, beyond which yields are unsatisfactory and the crop is not important in the economy; and the absolute limit at which the crop will grow and yield at all. Generally the amount of heat received is the most decisive factor in setting limits beyond which certain crops cannot reasonably be grown. Most of the other factors of the physical environment can be controlled or altered considerably, for example, water deficiencies can be overcome by irrigation.

The crop differences one finds from valley bottom to valley sides are significantly due to climate factors only between 1,800 and 1,300 m where sharp zonations, apparently thermally caused, are apparent. In the upper valley, such local differences are more because valley floors can be irrigated while the slopes are not. Nor are temperature inversions of much importance. While temperature inversions occasionally occur in the valley, they are not frequent enough to significantly affect the distribution of crops and there is no noticeable pattern of the more fragile crops being grown on the slopes with the hardier ones on the valley floor. Slope exposure is of some consequence in parts of the valley: below 1,000 m coffee and coca are not grown on north-facing slopes, while above 3,000 m north-facing slopes are more intensively cultivated than the south-facing slopes. Plant diseases and insect pests may restrict or even eliminate certain crops. Thus the potato does not grow successfully below 2,400 m in the valley, not because tubers do not form on the roots, but because fungus infestations destroy them. However the physical environment alone does not explain the distribution and specific combinations of useful plants grown in the Vilcanota Valley. From the total array of possible crops available to them, the inhabitants select crops which have proved to be most acceptable on the basis of physical, cultural and economic considerations.

While each human-manipulated plant has a particular ecological niche in which it grows and yields best, the range of its cultivation is wider in this peasant society than it would be in a modern industrial-type agriculture. The cash crops quite clearly occupy their favorable ecological niches, but those grown mainly for subsistence are cultivated over as wide an area as peasants can expect to receive some return, even if small. Thus coca and coffee both are grown from below 730 m (beyond the study area) to approximately 2,000 m; however, coca yields best, qualitatively and quantitatively, between 1,000 and 1,500 m, and coffee between 1,000 and 1,800 m, so most attention and land are given these crops in this zone. Certain commercial specialty crops also have a fairly limited range, such as maway potatoes and miska maize, both grown around Urubamba (about 2,800 m) where they mature early and thus can command higher, early-season prices. On the other hand, tarwi, primarily a subsistence crop, clearly exceeds its ecological niche in the valley. This plant is grown from 2,630 m to 3,830 m although SAUER places it in the 'cold lands' (1952:539). Such discrepancies are characteristic of the peasant subsistence agriculture in the valley in which a diversity of crops is desirable to obtain a greater variety of foodstuffs, as well as to assure that, if several crops are planted, at least some are bound to succeed.

By this means a greater variety of foodstuffs are available to the family – raising crops which mature at different times of the year. Therefore food supplies are extended through a longer period, and if several crops having varying requirements and enemies are planted, at least some are bound to succeed. Food preferences, traditional trade ties and simply local or family custom also help explain the wide distribution of or peculiar local specialization in one crop or another. All of these change but slowly and once a crop becomes established in a place, its maintenance becomes a matter of tradition and successive generations tend to follow the tradition. Only repeated disaster due to drought or disease or a notable economic change will alter the pattern.

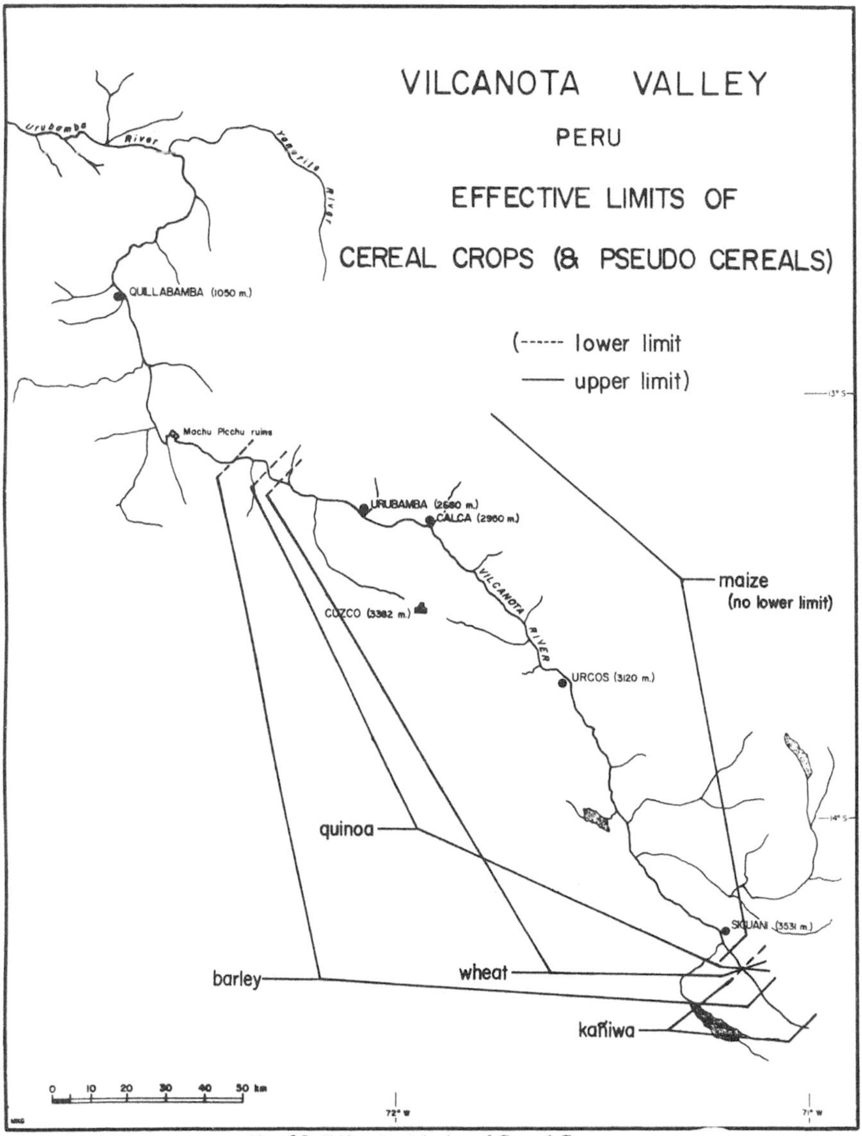

Fig. 38. Effective Limits of Cereal Crops.

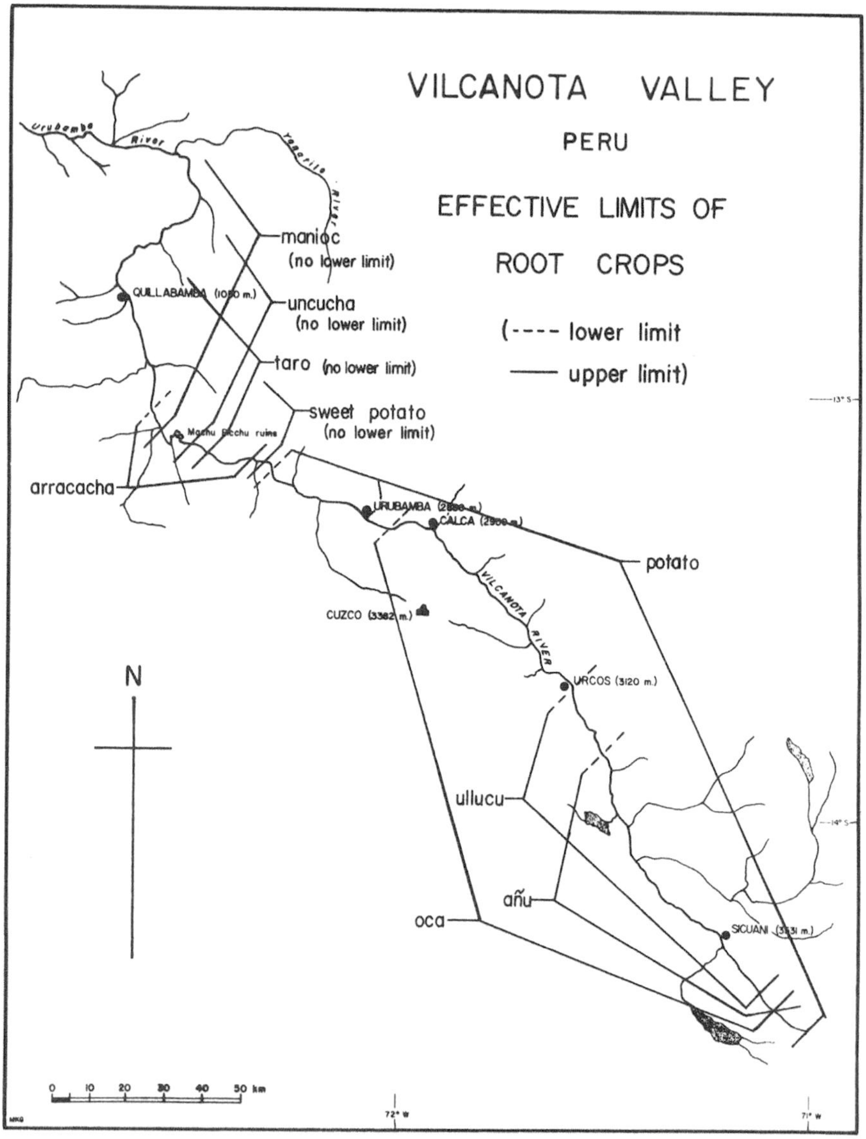

Fig. 39. Effective Limits of Root Crops.

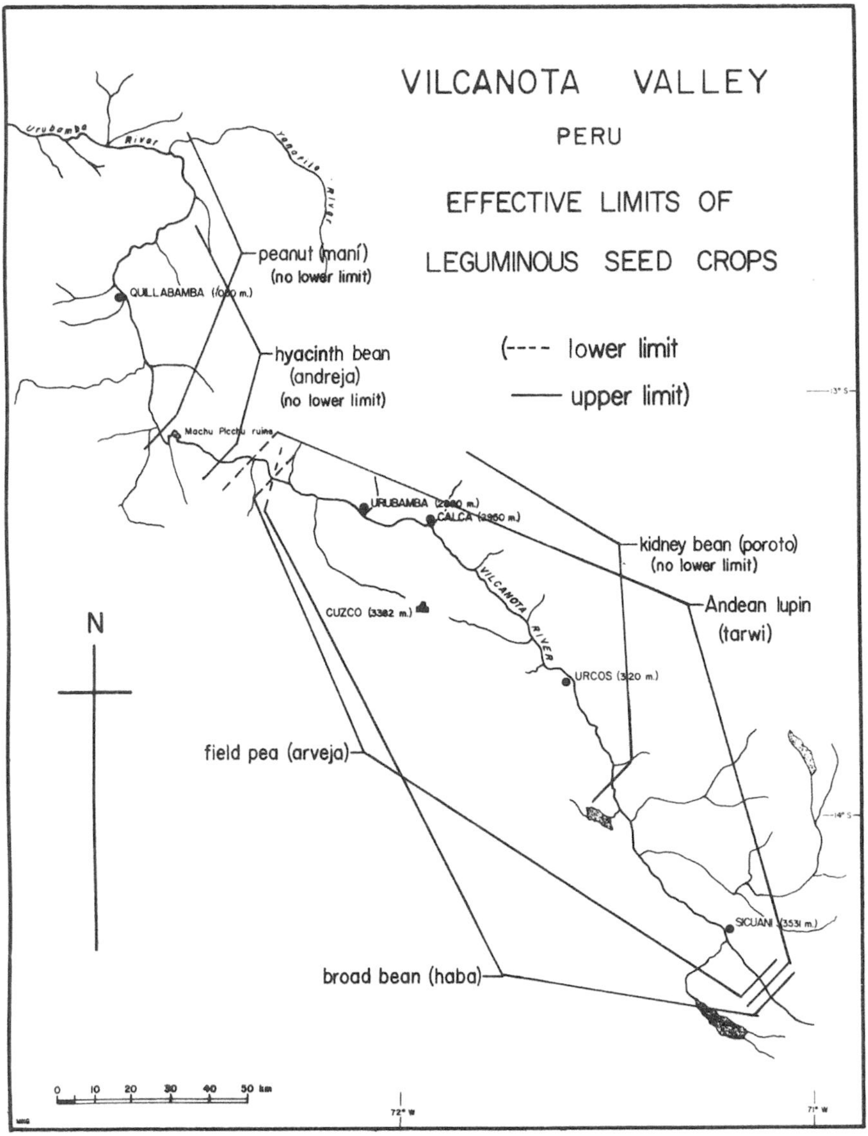

Fig. 40. Effective Limits of Leguminous Seed Crops.

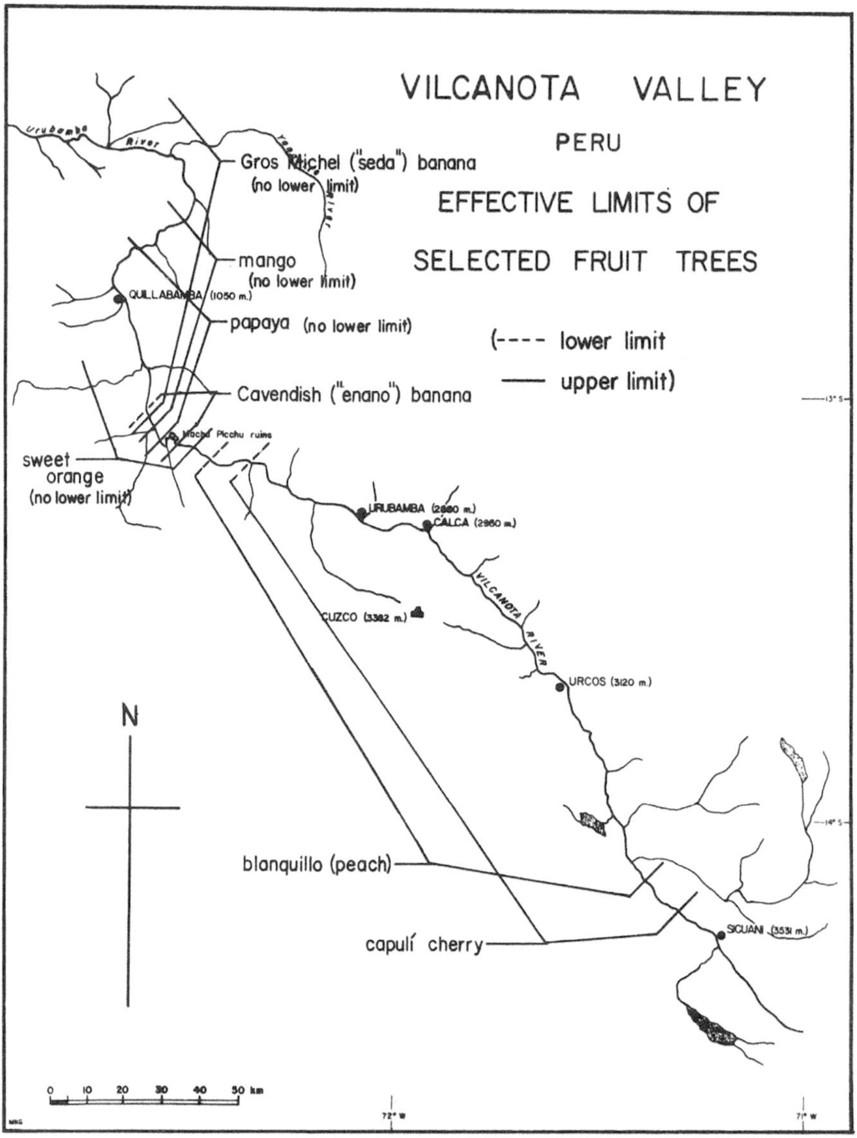

Fig. 41. Effective Limits of Selected Fruit Trees.

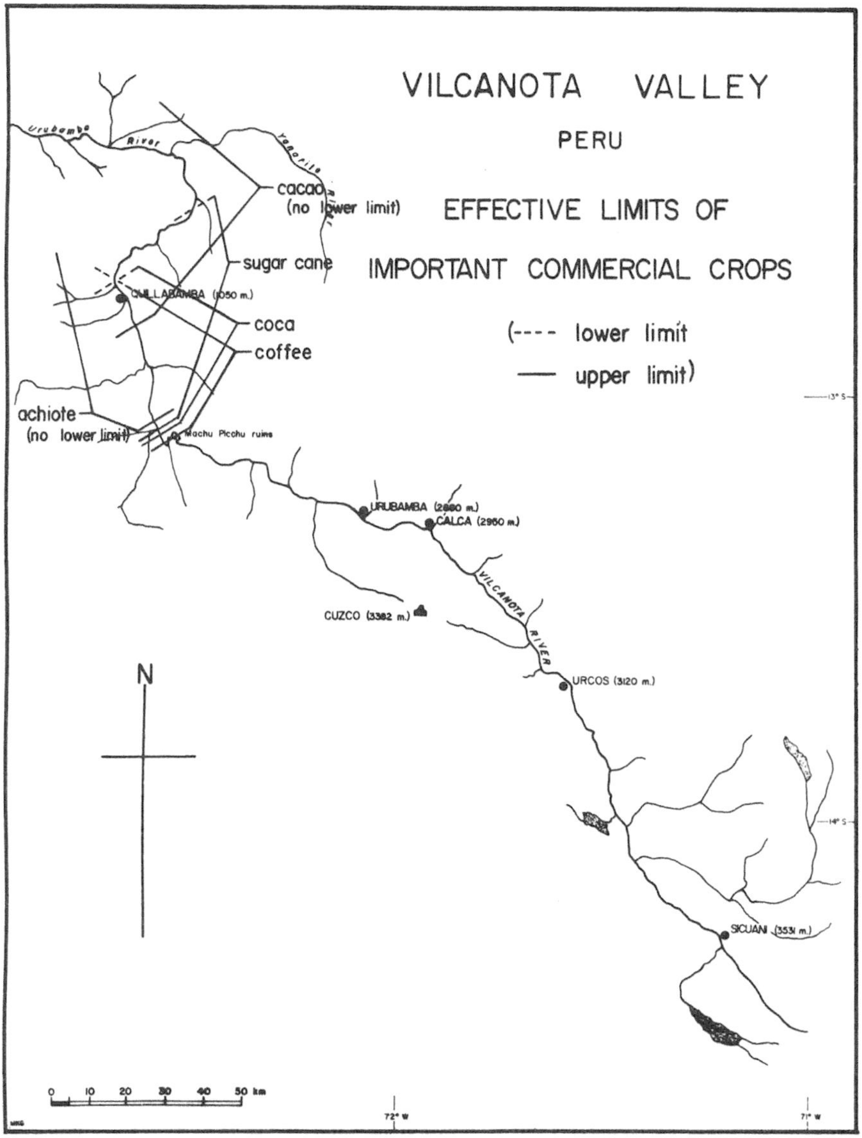

Fig. 42. Effective Limits of Important Commercial Crops.

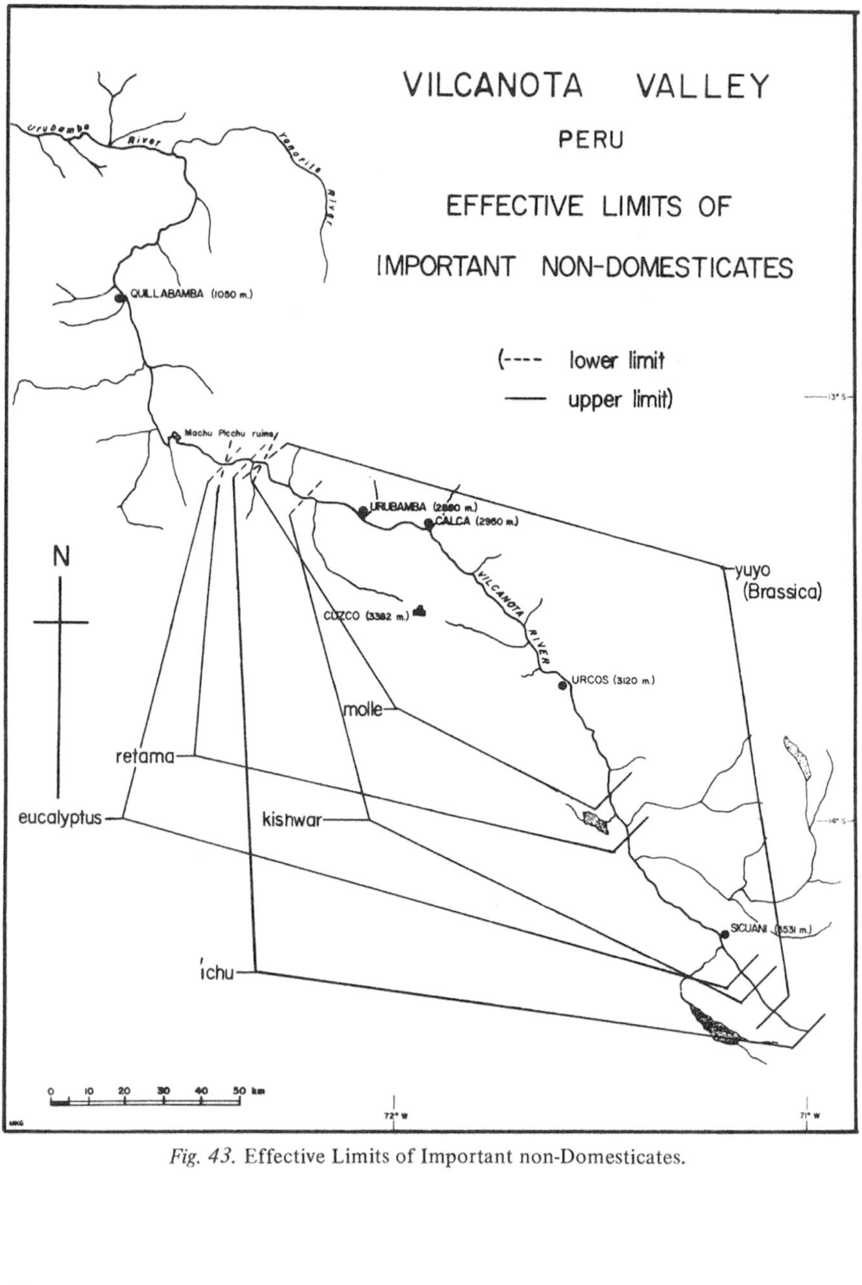

Fig. 43. Effective Limits of Important non-Domesticates.

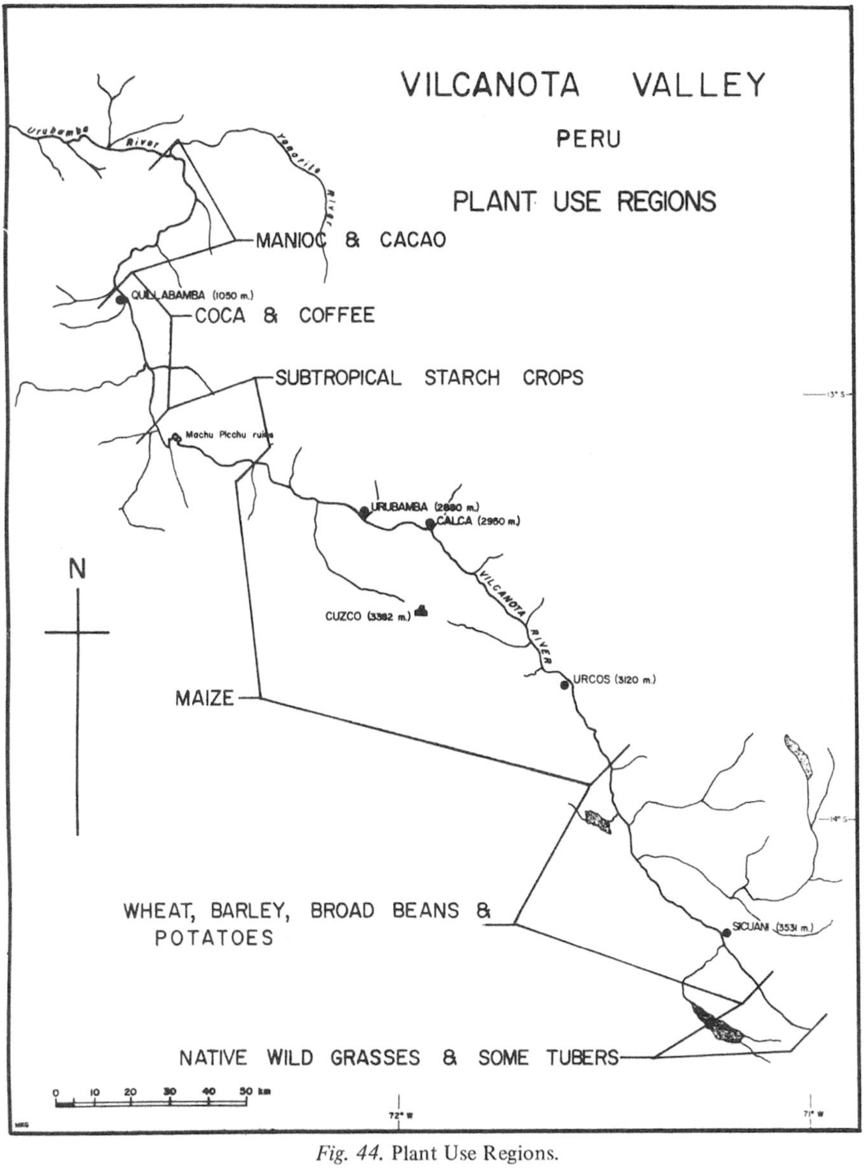

Fig. 44. Plant Use Regions.

The comparative distribution of important useful plants in the valley is shown in the following series of maps (Figs. 38-43), based on the effective rather than the absolute (or ultimate spatial) limits at which each is grown. Thirty-six species' ranges are shown, six of them are not domesticates but are important in the folk economy of the valley. The plants are grouped according to plant parts used (root crops, cereals and pseudo-cereals), botanical affinity (legumes), economic orientation (outstanding commercial crops) and degree of human manipulation (non-domesticates). The plant ranges within each of these groups show considerable overlapping, although they may occupy different niches in the same zone, that is on upper slopes, on lower slopes or in the irrigated valley bottoms. Taken together, this series of maps shows clearly that in the area between about 1,800 and 2,100 m the limits of many heat-tolerant and cold-sensitive plants on the one hand, and cool-tolerant and heat-sensitive plants on the other, overlap. Another locality of rapid crop change is south of Maranganí, upstream from 3,600 m where frost-sensitive plants are generally eliminated.

From another point of view, the important useful plants of the Vilcanota Valley may be divided according to the areas in which they grow. In this manner, one may define six ecological regions, each of which is characterized by a unique complex of useful plants which are very important in the livelihood of the people in that region (Fig. 44). These regions are 1. region of native wild grass and occasionally cultivated tubers; 2. region of wheat, barley, broad beans and potatoes; 3. maize region; 4. region of subtropical starch crops; 5. region of coca and coffee; and 6. region of manioc and cacao. Each of these six zones corresponds to the Andean landscapes of Inca tradition (PULGAR VIDAL, 1946) still used by Quechua-speaking Indians today: 1. Puna; 2. Suni; 3. Quechua; 4. Chaupiyunga; 5. Yunga; 6. Rupa Rupa.

Between 4,340 m and 3,910 m in the Vilcanota Valley is the Puna landscape, an area of predominant livestock grazing on the wild grasses with occasional tuber cultivation. Undoubtedly the most important single grass species is *Stipa ichu*, the basic food of llamas and alpacas, as well as a plant used directly by man in a number of ways. The low temperatures and short growing season rule out full-scale agriculture, and within this zone one finds the uppermost limits of crop cultivation.

The vertical limits of agriculture may be either the 'effective' limit or the 'absolute' limit. In the former case, the definition implies the upper boundary of the agricultural region in continuous fields. In the Vilcanota Valley, this is best defined as the zone in which certain hardy crops are well enough adjusted to the prevailing climate to yield satisfactory at least 3 out of 5 years. The effective limit is located at 3,910 m about 3 kilometers up valley from the community of Occobamba. In the 1964 harvest season, the fields held *común* barley, *marquis* wheat, broad beans, añu and potatoes. In that year all the seed crops were shriveled and small, while in the previous harvest year (1963) frosts killed the wheat, barley and broad beans, but spared the root crops.

The absolute limit of agriculture is the uppermost line of tilled islands beyond and above the continuously farmed zone. The absolute limit in the Vilcanota Valley extends to the lower slopes (4,340 m) of the valley at the very top of La

Raya Pass (4,314 m).[15] The crops included potatoes growing in several tiny fields and kañiwa growing in a former llama enclosure. Between the farm limit of 3,910 m and the field limit of 4,340 m, agriculture is distinctly peripheral to livestock herding of sheep, cattle, llamas and alpacas.[16] Because the Indians in this locality depend largely on their animals, the loss of their pathetically low-yielding crops due to frosts or hail would not constitute a catastrophe such as would crop failure below 3,910 m where animals are less important and people depend primarily on crop harvests.

This area between 3,910 m and 4,340 m may have been more completely cultivated in the past than it is today. BINGHAM (1923:123) in 1915 noted fields of kañiwa and potatoes at La Raya Pass also, which indicates that the 1964 field observation was not unique to that year. There is a distinct possibility that this Vilcanota puna was cultivated in Inca times, for it is certain that the Incas farmed more land in the valley than the peasants do today; whether this reflected the need for more food for a larger population, or the efficient organization of Inca agriculture is not at all clear. The bases of agriculture in the puna would have been bitter potatoes (for chuño), kañiwa, añu, ullucu, oca and perhaps certain hardy cultivars of quinoa. In colonial times, this area which lies between the effective and absolute agricultural limits may have been cultivated because La Raya Pass was an important stopping place on the route from Cuzco to Potosi. Much barley was planted as fodder for the horses, and tubers were planted to feed the wayfarers. How long this continued is not certain but an early nineteenth century French traveler wrote in reference to La Raya that '... the sun produces only a bit of barley which never ripens and which is fed to livestock as forage; also potatoes, yucas (*sic*)... The Indians also feed themselves on other roots' (BOTMILIAU, 1850:908). Perhaps the demise of this section as an important crop producing area began when the railroad and later automobiles obviated the need to tarry at La Raya tambo in order to provide sustenance for man and his beasts.[17] Today this cold zone is used mainly by llamas and alpacas feeding on the natural grasses of the area and a few shepherds trying their luck at growing the hardiest crops known to them.

In the region of wheat, barley, broad beans and potatoes (3,910 m to 3,300 m), three Old World crops and one New World tuber predominate. These plants are generally hardy and often resistant to light frost. In the Inca period, this zone (known as the Suni) was undoubtedly characterized by tuber cultivations as well as by native seed crops. Today oca, añu, ullucu, quinoa and tarwi are of very minor

[15] According to KOVALINSKY (1934:274), the uppermost limit of crop plants in Southern Peru is at 4,500 m where potatoes are cultivated. WEBERBAUER (1945:626) also found potatoes as the highest crop, but at the somewhat lower elevation of 4,200 m.

[16] The large government alpaca farm which occupies much land (mostly in natural pasture) between Aguas Calientes and La Raya Pass practices a little cultivation of *kompis* potatoes, forage oats (for their horses) and a hardy kind of alfalfa. As it is grown without any serious risk to the growers, this example is not part of the folk agriculture and can be disregarded in considering crop limits.

[17] Aerial photographs taken in 1954 show faint lines of what appear to be lazy beds on most of the lower slopes from 3,900 m to 4,300 m, most of which are not cultivated today.

significance in the agricultural and domestic economy. The most important non-domesticate is eucalyptus, a rather late Old World introduction.

In no other section of the Vilcanota Valley is so much land or agronomic attention devoted to one crop as in the maize region (3,300 m to 2,400 m). Maize is overwhelmingly the leading crop and occupies perhaps 90 percent of the irrigated land in this zone. If irrigation water were not plentiful, maize would be less important. Only on the unirrigated slopes above the elevation of Ollantaytambo are small grains cultivated. Rosaceous fruit trees (pear, peach, plum, etc.) abound from Calca to several kilometers below Urubamba. Eucalyptus is, as in the preceding zone, the most useful and conspicuous non-domesticated plant, while on the roadsides retama (*Spartium junceum*) is everywhere.

Almost all of the maize here is of the flour type which thrives in the rather coolish temperatures prevalent in this part of the valley. Abundant irrigation water is provided by the copious meltwaters entering the valley from glaciers in the Vilcanota Range. Basically this zone has probably changed the least from the Inca period when these Quechua (or Keshwa) areas were favored above all others for human settlement and agriculture. Assuming an early introduction, maize has dominated the agricultural scene for at least 1000 years. The marked concentration on maize in this section of the Vilcanota Valley is in sharp contrast to the corresponding section of the Paucartambo River Valley, the next major north-south depression to the east, where the specialization is the potato. These differences cannot be explained by significant differences in the physical geography of the two areas nor by the present-day economic geography of the Department of Cuzco. One might speculate that the present crop patterns in the matching parts of these two valleys may have been conditioned by historical prescriptions by which the crop to be grown was specified by Inca law or governmental design in keeping with a larger regional plan.

In the region of subtropical starch crops (2,400 m to 1,500 m) the main crop is arracacha, with sweet potato, dasheen and uncucha being much less important. In the lower part of this zone, the Cavendish banana is important. Nevertheless crop diversity is an outstanding fact of the region. The zone is the Chaupiyunga of the Incas who probably found it no more attractive for settlement than do most of the Highland Indians today. Cultivation of the steep slopes is limited by the high rainfall and the small amount of land on which crops can be grown. The zone forms a clear buffer between the tropical and subtropical types of crops. The wild flora in the zone provides most construction materials, combustibles and medicinals.

The region of coca and coffee extends from 1,500 m to 1,000 m. In Inca times the cultivation of coca was the prime reason for occupying the Yunga. In the colonial period, sugar cane was introduced and remains on the scene today, but its importance has declined in favor of coffee. Whereas cane provided a source of income for the large estates, coffee is grown by small peasants and estate owners alike. Coca is still very important but its future is problematical, and coffee appears destined to fill the ecological gap that may eventually be left by it. Fruits, particularly the orange and banana, also are largely for commercial sale, and manioc and maize are grown for subsistence purposes. The wild flora furnishes construction materials, firewood and medicinals.

The region of manioc and cacao (1,000 m to 730 m) is considerably less commercially-oriented than the preceding one because it is farther from the railroad and the Highland market and subsistence crops receive more attention. Manioc is the dominant chacra crop and the dietary base; little is traded. Some coca and estate-produced coffee come from this zone, but the main commercial crop is cacao. The use of wild flora is similar to that in the preceding zone. What the pattern was in Inca times is not known, and one can only guess that this zone, partly in the Rupa Rupa, may have been an area of contact between Highland Quechua-speakers and the forest Indians. As such, it may have been the threshold of introduction for many lowland material culture traits from the Eastern Lowlands.

3. PAST AND PRESENT USE OF THE GREEN MANTLE

The peasants in the Vilcanota Valley have at least some use for almost all of the phanerogamic plants that grow in that depression. Because most plants have value to them, and they live closer to them, these people have a much keener knowledge and appreciation of plant names, uses and folklore than do individuals in industrialized, urban societies. Generally, Indians know more about plants than do cholos or mestizos and peasants at higher elevations know more than the people at lower elevations, many of whom are originally from the Highlands and thus somewhat out of their native environment.

Not only have the peasants adopted the Spanish names for all Old World plants, but they use Hispanicized versions of their own native plants with increasing frequency: wheat is *trigo*, sugar cane is *caña* (*de azucar*), potato is *papa* (Quechua name lost), and sweet potato is *camote* (*apichu/kumara* in Quechua). But most indigenous South American crops and wild plants have local names which are still used, because Spanish equivalents never took hold among the Spaniards whose lexicographical habits have been followed by the Quechua speakers. Thus edible canna is *achira*, the two domesticated chenopods are *quinoa* and *kañiwa*. Other native plants have both Quechua and Spanish names (or names which, while not truly Spanish in origin, were adopted and diffused by the Spaniards) and which are used with equal frequency among the peasants: *yuca* and *rumu* for manioc, and *lisas* and *ullucu* for *Ullucus tuberosum*. Still some names for native plants appear to be completely lost: annatto is now always *achiote* rather than the Quechua *mantur*; tobacco is no longer *sayri* but *tobaco*; *sauce* (willow) has replaced *wayaco*; *piña* (pineapple) has replaced *achupalla*; and *granadilla* (passion fruit) has replaced *tintin*.

To indicate variation within a species the indigenous peoples long ago devised a useful taxonomic system which has little in common with scientific classification. A cultivar of a particular cultivated plant is named for the most specific useful characteristic of the valued part of the plant. Potato tubers, for example, are named for one or several of the following: color, shape, cooking quality, and environment in which it grows. Description to differentiate potatoes based on such characteristics as the shape and color of the corolla or the growth habit of the plant itself or the number of chromosomes in the cells which indicate the evolutionary level of the potato is not known.

The soundest procedure in discussing useful plants is to follow a taxonomic organization because the useful plants of a folk society have no neat ecological niche or special distributional zone into which they can be pigeonholed, nor have they a single purpose or function. For example, it is a vast oversimplification to consider quinoa under the rubric of 'plants used for their seeds', for quinoa is virtually as important for its edible leaves and the alkaline ash made from the stalk as it is for its edible seeds. Furthermore, a discussion of useful plants organized taxonomically helps to convey to the social scientist an appreciation of the objective reality of the plant and serves as a constant reminder that all crops and other plants modified by man had their origin in the wild flora. If one is to understand the process of plant domestication, an understanding of taxonomic relationship is essential.

The classification of families used in the following discussions follows the Engler-Prantl system.[18] Most useful plants in the valley are angiosperms (the more highly evolved plants with seeds enclosed in an ovary), a not unexpected finding since probably 95 percent of the valley flora belongs to this group. Unlike in the North American and European highland areas, very few conifers occur naturally in the Andes. Certain families of plants are especially important ethnobotanically in the Vilcanota Valley. These are, in approximate order of usefulness, quantitatively and qualitatively, to man: 1. *Gramineae* (grass family); 2. *Leguminoseae* (pea family); 3. *Solanaceae* (potato family); 4. *Umbelliferae* (celery family); and 5. *Compositae* (sunflower family).

Cultural-Geographical Analysis of the Individual Floristic Components: Plants other than Dicots

NON-ANGIOSPERMS

Nostoc sp., *llullucha* or *kochayuyo* is a gelatinous, dark green alga that grows wild

18 ADOLF ENGLER & KARL PRANTL published Die naturlichen Pflanzenfamilien in 1909 which contained a taxonomic system followed in nearly all manuals and floras in most herbaria. The order of genera discussed within each family follows the organization of DARLINGTON & WYLIE in Chromosome Atlas of Flowering Plants, 1956.

Only those plants which I considered to be of some economic and/or cultural importance are included in the following floristic discussions, although to be sure, almost every plant in this peasant society is 'used' in some way. Approximately 60 percent of the total number of plants in the upper valley are discussed, and 20 percent of the flora in the lower valley.

I am most concerned with plants at the species level in this Chapter 3. The variety, as a subdivision of the species, is introduced when appropriate, for much intra-specific variation, especially of crops, has relevance in cultural geography. In this study, variety will be considered in the agronomic sense and not the botanical sense of the word, and will be used interchangeably with the term 'cultivar'. The Random House Dictionary of the English Language defines 'cultivar as a variety of plant that has been produced only under cultivation. The variety of cultivar as here used is a floristic unit possessing certain morphologic and/or economic and/or agronomic individuality which separates it from other units of the same species. No claim is made that the intra-specific variation in the valley discussed for each crop is complete. Vouchers of many of the wild and cultivated specimens are available from the author's collection and from the University of Wisconsin Herbarium in Madison. Other identifications were made by Dr. CÉSAR VARGAS C., professor emeritus of botany at the Universidad Nacional de San Antonio Abad del Cuzco.

in lakes and ponds and is harvested during the dry season. COBO (I:179) mentioned it as a food in the early colonial period, and its use undoubtedly goes back to the Incas (GUAMAN POMA DE AYALA, 1966:III:131). Today it remains an Indian item of consumption and is seen on sale in markets after it has been dried and pressed into rectangular blocks. Llullucha is usually eaten with capsicum pepper in a stew or soup during the food shortage season, and it provides vitamins to Indians living at high elevations. Possibly its main value is tied to the fact that it contains iodine and therefore helps prevent goiter, a common deficiency disease in the valley.

Dryopteris palacea (Swartz) C. Christ, *raki raki*, is a wild fern collected in the valley below 2,600 m as a famine food. The young fronds (*hullpo*) contain starch and are eaten when still folded. The leaf is also used in infusion as a mild purge.

Cupressus sempervirens L., *ciprés*, is native to the Mediterranean Basin and was introduced into Peru in the sixteenth century (COBO: I:403). This evergreen ornamental occasionally is planted in plazas and other open spaces in the upper valley where the branches are sometimes used to make wreaths for the dead.

Ephedra americana H&B ex Willd., *pinko pinko*, is a wild shrub used to treat bladder trouble and inflammation of the gums.

GRAMINEAE

Maize

Zea mays L., *sara, maíz* or maize, represents such a triumph of environmental and cultural adaptation in the Vilcanota Valley that a local Peruvian origin is often presumed. It does not, however, now appear that South America was the place of domestication, for the accumulated archeological evidence points strongly to Mesoamerica as the original cradle of maize culture. The earliest dated appearance of primitive-type maize in Mesoamerica is *c*.4,000 B.C. (MANGELSDORF et al.: 1964) and South America *c*. 750 B.C. (TOWLE, 1961:22).

The relative pre-Conquest importance of maize in Sierra agriculture is difficult to determine because of the lack of archeological plant material in the Highlands. Maize, not tubers, was the principal ceremonial crop in most of Highland Peru and MURRA (1960) proposes that this was a manifestation of the Inca anxiety about the plant's precarious place in the environment and/or economy. Several festivals in the Inca calendar were dedicated to maize (GUAMAN POMA, 1956): in January, *Hatun Pokoy* was celebrated to help make the maize grow well; in April, *Airiwa* was the maize ripening festival; and in May, *Aymuray* celebrated the maize harvest. Words may also provide a clue, for the Quechua language lacks a full vocabulary for maize and its different stages of growth, a notable contrast to the rich terminology for the plant in the Tarascan language of Mexico (WEST, 1948).

Although maize originated as a crop in Mesoamerica, Peru was undoubtedly a major center of maize evolution. After its diffusion southward, maize gradually evolved into different types aided by its ease of cross pollination. Different genetic strains then developed in isolated river valleys where there was little subsequent seed exchange. Occasionally hybridization may have occurred by chance from some outside seed material. This may have resulted in a vigorous new kind of maize which had advantages over the old types and was therefore preserved. But since at least the Conquest, four centuries of cross breeding have resulted in heterozygous maize both in its gene composition (genotype) and in its appearance (phenotype) and today xenia (i.e. kernels of different shape and color in the same ear) is more the rule than the exception in peasant-grown corn. One can assume that the reservoir of recessive genes is enormous. As long as different cultivars are planted together in the same fields, contamination will take place because the plants flower about the same time and open pollination prevails.

Modern hybrid maize has not yet entered the valley as a crop. Some agronomists claim that in the Vilcanota Valley, the maize yields could be raised from the present 3 tons per hectare to over 9 tons per hectare if select hybrid seed were used. In spite of this impressive increase in yields, the use of a single hybrid maize cultivar to the exclusion of others, would be of doubtful benefits to the peasants of the valley. Such a change would require them to prepare different maize dishes by methods to which they are unaccustomed and would necessitate far more discipline in order to perpetuate hybrid generations than the peasants are prepared to follow.

Maize Classification

Maize is classified by the inhabitants of the valley according to provenance, quality, kernel size, induration or color pattern. The terms used to classify maize according to its source or where it is grown are relative designations depending on one's location. In the upper valley, maize of the local area is simply *sara*, whereas maize from outside the area is *wawa sara*. In this case, the local maize happens to be floury, and the outside maize is flint or dent type from the hot and fairly wet lower valley. On the other hand, the Machiguenga Indians of the Urubamba Valley (just below the study area), use the word *chinki* for local flint maize and *tancoki* (meaning 'outside maize') for the flour maize brought down from the Highlands by trade.

After the harvest, maize is also classified according to its quality. A basic distinction is made between *allin sara* (good maize) and *huttu sara* (rotten maize). The former is sold or is used for domestic consumption, while rotten maize, if it is completely moldy, is fed to pigs; if infected with *Ustilago* fungus, the peasants eat it. If the maize is still in the early stages of rot, it is made into chicha by being milled together with the sprouted maize (*jora*).

From the portion of the maize harvest destined for sale, those ears having mostly large kernels are set aside before shelling as 'first class' (*primera*), while ears of smaller kernels are 'second class' (*segunda*). Maize with very small kernels, known as *hanokchi*, has little commercial value and is kept by the people for domestic use or, in the case of the hacendado, is given to his peons. Where peasants grow maize entirely for their own domestic use, the maize is not sorted according to first, second and third class. Instead, the kernels of maize are sorted after the ears have been shelled according to size (the actual dimensions of which are relative to corn size in the local area). Large maize kernels (*hatun sara*) are used for boiled corn (*mote*), while the small kernels found near the end of the ear (*kaminko sara*) are preferred parched or milled.

As a generalization, maize grown at very high elevations or at very low elevations, tends to have hard kernels, and unirrigated maize is harder than maize grown under irrigation. An essential distinction that goes back to Inca times is the classification of maize into 'hard' and 'soft' types. GARCILASO DE LA VEGA (1960:36) noted two kinds of maize in Cuzco: *muruchu* (or *morocho*), a hard flinty maize, and *capia*, a soft presumably floury type of maize. COBO (I:160) used *morocho* to designate the hard maize of the lowlands that the Spaniards fed to their horses, and this term is still current.

Maize is differentiated on the basis of its color or color pattern. The main solid pigments of the pericarp are, in approximate order of occurrence, the following; yellow, white, dark red, dark grey, brown and black. A red and white variegated pattern and a mottled purple and white pattern are the two most frequently seen maize ears with a non-solid color.

Color pattern, induration, and kernel are the main criteria by which the peasants segregate maize into distinct cultivars. The scientific criteria which are currently used in classification, such as width and length of leaf, tassel length, cob diameter, chromosome knobs and B-chromosomes, are not relevant to the peasants.[19] Some maize cultivars fit fairly well into the accepted subspecific designations of flour corn, sweet corn, flint corn, dent corn and pop corn. In addition, however, some maize does not fall neatly into any one of these types for it combines the characteristics of two. Most of the maize grown above 2,400 m is the flour type or a type intermediate between flour and flint maize. This corn is important because it grows efficiently in an environment having temperatures which are relatively low for maize and because of the food preferences of the indigenous population. The lower valley limit (about 2,400 m) of the floury type maize is due to the high humidity which causes the corn to rot, and the upper limit (about 3,600 m) is set by low night temperatures. This maize has large and often slightly pointed kernels, eight to ten rows on a cob, a variety of kernel colorings and a high starch content compared with other kinds of corn.[20]

19 The classification of Peruvian maize into 50 races by GROBMAN (1961) and his co-workers at the La Molina Agricultural Experiment Station is the outstanding contribution to the scientific study of maize in the country. Of the 50 races defined, 16 were in the Vilcanota Depression, but not all of these are recognized ethnographically.

20 Four hundred and twenty dried ears from the Vilcanota Valley were collected in the 1963 and 1964 harvest seasons. This maize came from three sources: directly from the field, from the grain bin and from the public market; therefore, it does not represent a random sample. This collection, in the author's possession, permitted a close examination of the variation in the maize in this valley.

Fig. 45. Maize cultivars from the Urquillos area around 2,950 m. From left to right the varieties are *yana* (used for dyeing); *chile* (used for popping); *chullpi* (used for parching); *parakay* (used for boiling); and *kulli* (used for chicha).

A. *Parakay Sara* This cultivar, also called *blanco* or *Cuzco*, has achieved considerable fame even beyond the borders of Peru, for its kernels are perhaps the largest known anywhere. The ear is not particularly long (about 12 centimeters) but it is notably thick and has eight rows of kernels (Fig. 45). Both the kernels as well as the cob are white. The plant has large leaves and attains an average height of almost 2 meters. Parakay requires 8 to 9 months to mature, presumably in order to develop the high starch content. This vegetative cycle is longer than that of other varieties and the Quechua name, which means 'early', must refer to the fact that parakay must be planted early, rather than to an early harvest time.

Parakay is grown in the valley between 3,300 m and 2,800 m and may have originated here as a distinct cultivar; at least this area is recognized as its most favorable environment for growth. The fourteenth and fifteenth century Incas may have been responsible for its dispersal and for its present distribution from Ecuador to northern Argentina. Hacendados and small mestizo farmers are responsible for most of the parakay production and they grow it in fields by itself separated from other cultivars. When peasants raise it, the result is a phenotypically impure maize because they mix it with other cultivars and much cross-breeding occurs. This maize has a high water content and takes at least 25 sunny days to thoroughly dry. It is appreciated as green corn (*choklo*), boiled corn (*mote*) and for tamales (*humintas*), but it is not acceptable for chicha which may explain why peasants do not grow it in large quantity.

Blanco imperial, as the hacendados call it, is an extremely large-kerneled variant of parakay which does not qualify as a truly separate cultivar among the peasants. GROBMAN et al. (1961:299) consider this giant-kernel kind as a highly evolved separate race derived from parakay either by hybridization with another cultivar or by a benign mutation capitalized on by Inca farmers. It is grown between Coya (2,995 m) and Hacienda Tarabamba (2,830 m) but its large-scale production is centered on the area between the town of Urubamba and Calca. The kernel size is to some degree the result of heavy fertilization and only hacendados produce this enormous size. Yields may be as high as 1,900 kilograms per hectare.

B. *Uwina Sara* Also called *amarillo*, this is the most popular maize cultivar above 2,500 m in the valley. The ears of this maize are typically conical, short (13-14 centimeters) and thick with large, rather flinty yellow kernels arranged in eight regular rows. In contrast to parakay, the cob is hard, relatively thin and reddish on the exterior. Uwina is sown in October and is ready to be harvested in about 7 months. It is cultivated throughout most of the upper valley, but it is most important, relatively, between Combapata (3,465 m) and Piscocucho (2,635 m) and it yields best in the Ollantaytambo area. Peasants prefer uwina over parakay maize because it has a wider variety of food and drink uses: it is used for chicha, in boiled form and also toasted.

Several other cultivars are grown on a smaller scale than parakay and uwina and have been selected for their special characteristics of color and flavor. They include the following:

C. *Saksa* This cultivar has blood-red streaks which run through the middle of each individual kernel. The kernels themselves may be yellow, white, pink, or orange, and are often beaked. Saksa is cultivated widely in Calca and Urubamba Provinces and it used mainly toasted.

D. *Granada* This floury-flinty maize has small ears used boiled and as meal. The kernels are pinkish at the base, but otherwise are white. Granada is perhaps a hybrid of parakay and kulli. It is grown especially in Qiuspicanchis and Canchis.

E. *Kosñisara*. This an uncommon color variety of yellow flour corn with brown kernels; the first part of its name means 'smoke' in Quechua. It is found from about San Pablo (3,490 m) to Ollantaytambo (2,790 m) and consumed mainly boiled.

F. *Chaminko* This is a variegated red and yellow flinty-floury cultivar with long ears. The peasants appreciate it for making chicha.

G. *Kulli* Also known as *capulí*, this cultivar has reddish-purple kernels which appear to be almost black at times. The ears are short. It is a prime chicha maize and is grown from about 2,700 m to 3,300 m.

H. *Piskorunto* This name is a corruption of the word meaning 'bird's egg' in Quechua and refers to the mottled purple pattern of the kernels. The plant is short and precocious, and grown especially in Canchis and Quispicanchis. Among the floury kinds, it is said to have the highest sugar content, and the peasants prefer to toast it.

I. *Okkè* (or *Occe*) This is a small-eared corn cultivar with lead-colored kernels

Fig. 46. Two maize cultivars from 1,400 m; on the left is Cuban Yellow (*cubano*) and on the right is *chuncho*.

smaller than those of most floury maize. Found most commonly in Canchis and Quispicanchis, okke is used primarily for *kancha* and to a lesser degree *mote*.

J. *Yana* This is a very deep purple or black maize that is used as a dye rather than as a food. A small plant with small ears, it is planted on the margins of maize fields from Canchis to Ollantaytambo. (Fig. 45). At one time it may have been used to color ritual foods; today it is grown to tint chicha a purple color. It is one of the rarest cultivars grown in the valley.

K. *Chullpi* This yellow, long-kerneled cultivar has unusually shaped ears which, in some cases, are virtually round. Chullpi has a high sugar content, but it is not eaten immature, 'corn-on-the-cob' fashion as in North America. Instead this maize is dried and toasted to make kancha. It is consumed in no other form. Most peasants in Urubamba, Calca and Quispicanchis Provinces grow chullpi but on a small scale. This may be because it has limited uses as a food and/or its low yields. Only one ear forms on a plant except where the land has been highly fertilized. Chullpi is frequently planted in the middle of a field, surrounded by other cultivars of maize, in an attempt to conceal it from would-be thieves. It is no wonder, then, that chullpi is often impure and many ears are dotted with kernels of both uwina and parakay. If often takes on the longer cob structure of these two other maize cultivars (Fig. 45).

L. *Cuban Yellow* Also known as *cubano* or *nuevo*, this cultivar is a twentieth century introduction into the valley that is rapidly increasing in importance. The flinty kernels are a brillant orange-yellow, the cob is white and the ear is fairly long (Fig. 46). Although it can be grown in both the lower and upper valley, the cultivar is becoming widespread especially below 1,500 m because it effectively resists most insects and fungus diseases. Its yields are relatively high, and it is well suited for corn meal. The exact introduction date is not clear and, in fact, there may have been several. Taken to Cuba from Argentina between 1914 and 1930, Cuban yellow was introduced in Peru and released to farmers in 1948 (BROWN, 1960:40). An earlier introduction of this maize, however, was made in 1923 at the Evangelical Mission near Calca (GOHRING, 1934:32). Since most evidence in La Convención points to an introduction within the past 20 years, it is likely that the maize grown by the farmers today came originally from the 1948 introduction.

M. *Kallwa Sara* This little-grown cultivar is found in the valley above 2,800 m particularly in Calca and Quispicanchis. It is something between a flour maize and a flint maize. Kallwa has a smallish ear and the kernels are whitish with a certain transparency. Kallwa matures earlier than other cultivars in the same area (3-4 months to reach maturity, as opposed to 7-8 for others), and the ears and plant are smaller than those of the others. The size difference may be explained by the fact that it is cultivated on the unirrigated slopes of the valleys as a *temporal* crop, Kallwa is primarily made into flour or meal.

N. *Chile Sara* This is popcorn with very hard, small kernels which are always pointed and often beaked. The plant is short and the ears are very small (Fig. 45). Chile sara maize matures in about 5 months and is often grown on the periphery of fields of other kinds of maize. Because it flowers at a different time, there is little cross-breeding with other cultivars. Popcorn is grown in the Provinces of Urubamba, Calca and Quispicanchis and it is particularly common around the town of Urubamba. Cholas pop the corn for special occasions such as fiestas, but it is by no means a common, everyday food.

O. *Laurel* This is the main maize cultivar in the ceja de la montaña. It is grown from an elevation of 2,500 m to 1,300 m and may have been the maize grown at the Inca city of Machu Picchu. Laurel, a very tall maize, is sown in October and harvested in February. It has long ears, thick cobs and transparent hard kernels. It is used boiled, for making tamales and milled to be made into a mush.

P. *Pirisinko* This maize is grown at low elevations below 1,000 m in the valley. The long ears contain very small and hard yellow kernels which are used for meal and chicha.

Q. *Chuncho* This cultivar is a tropical dent corn with very long, cylindrical but truncated ears and an unusually thick cob (Fig. 46). Ordinarily there are 16 rows of kernels, usually yellow-range, sometimes with red streaks. Chuncho is a very tall growing maize which is cultivated today downvalley of 1,720 m on steep slopes without irrigation. It is sown most often in August on recently cleared land and harvested in November and December. A second crop is planted in December or January and harvested 4 months later. A third crop is possible only if the land is irrigated and since maize does not receive priority on such land, only two crops a year are common. Chuncho is the Quechua name for 'barbarian' which was applied to the forest Indians; this name suggests that perhaps the cultivar entered the

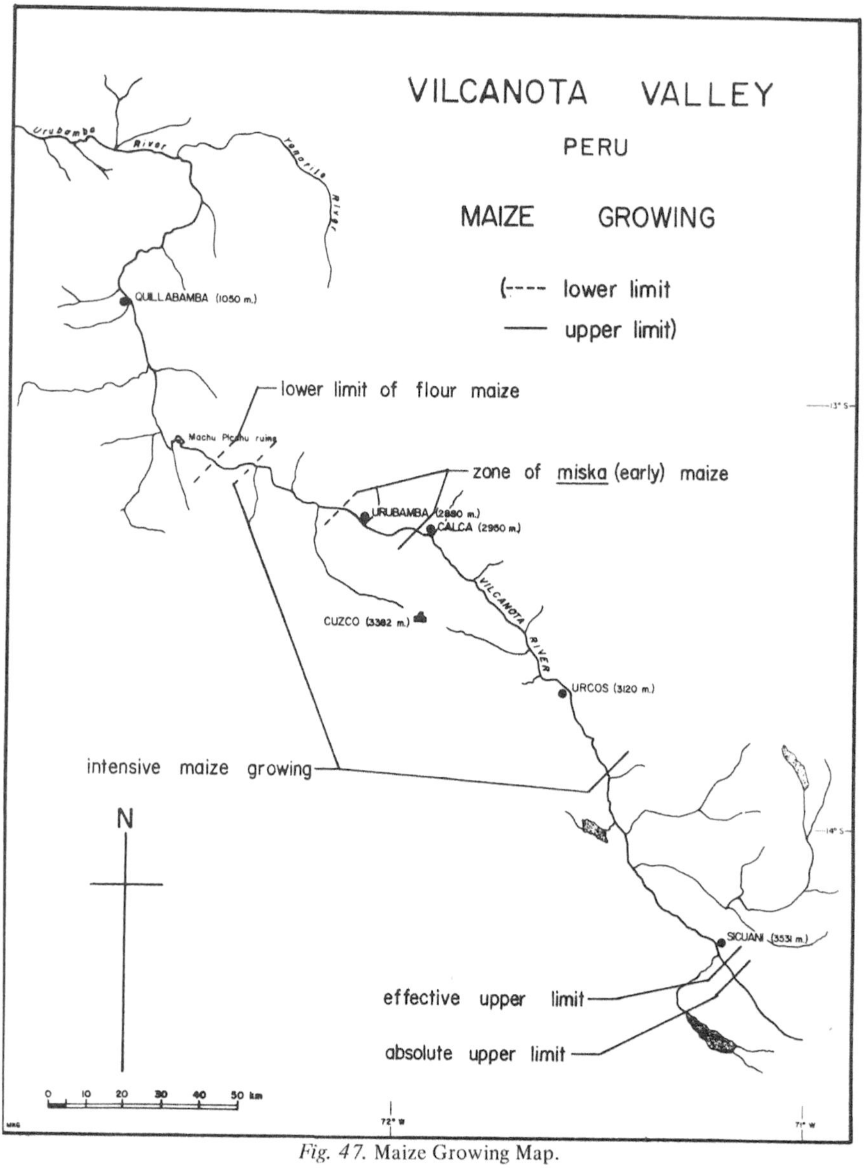

Fig. 47. Maize Growing Map.

valley by way of the Amazon drainage basin and was subsequently adopted by the Highland Indians as they moved down into the lower valley. Chuncho is milled and prepared as a mush. Some peasants do not care for the flavor or texture of this maize and use it only as chicken feed, a radical departure from the prevailing use of maize as exclusively human food.

Maize Distribution

Maize is cultivated in the Vilcanota Valley from about 3,600 m to 730 m and thus has a greater altitude range than any other crop (Fig. 47). Its uppermost limit in 1964 was at an elevation of 3,600 m, about 1 kilometer south of Maranganí; tiny plantings were made as high as Luracachi, 3,720 m, but in that year as in most, frost prevented the maize from yielding even nubbins. The repeated attempts to cultivate maize under almost impossible climatic conditions indicate the peasants' strong attraction to this crop which cannot be explained by the economics of its production. Between Maranganí and San Pablo (3,490 m), maize suffers from harsh climatic conditions; yields are small and many kernels are tiny, wrinkled and have little nutrient substances.

Maize in Canchis above 3,465 m, near Combapata, is highly mixed gentically; some 80 percent shows introgression (i.e. genetic mixing) of cultivars, and thus is difficult to classify. Estate owners in this part of the valley do not grow maize commercially and its production is largely in the hands of the peasants who plant the seeds of several different cultivars in the same field. The maize plants are about $1\frac{1}{2}$ meters high, ears are typically short (10-12 centimeters), the cob is thin and kernels are small with a tendency towards beaking. The kernels also tend to be widely spaced rather than compact on the ear and the bare cob shows between rows. Although basically flour corn, the kernels tend to be flinty and much harder than the flour corn from Urubamba or Calca. The position of the ear on the stalk is very low; in a field south of Sicuani, the ears measured only 12 centimeters above ground level and were growing on plants a little shorter than $1\frac{1}{2}$ meters high.[21] In addition, this high altitude maize has an abundant adventitious root development which is not apparent on maize at elevations below 1,800 m.

In most of Canchis, maize is a hazardous crop to grow because of the possibility of unseasonal frosts. The decision to grow maize here is related to its high value as an item to be traded for meat and cheese from the high province of Canas. In 1968, a severe January frost killed most of the maize in Canchis, a loss the farmers bore with remarkable stoicism. Only below Combapata (3,465 m) can one be assured of a harvest in 9 out of 10 years on the average. In the Quispicanchis section of the

[21] It is highly possible that a statistical study would reveal a displacement of the maize ear up the stalk with lower elevations in the valley and thus higher temperature values. At Calca, the maize ear of 10 samples averaged out to be 76 centimeters from the ground, and at Quillabamba (1,080 m) 112 centimeters from the ground. Much more work would be necessary to verify this observation.

valley, the climatic conditions are much more favorable. Maize here grows well on the valley bottom and the majority of irrigated land in this province is devoted to its cultivation. Many cultivars are grown, including piskorunto (a favorite around Quiquijana), parakay, uwina, kosñisara, okke, chile and chullpi. Near Andahuaylillas (*c.*3,100 m), one approaches the upper limit of the full range of maize cultivars in the upper valley.

In Urubamba and Calca, maize is grown on approximately 85 percent of the irrigated land. The richest sites are the river terraces near the Vilcanota, but floods occasionally destroy the crop growing on them. Uwina is the outstanding cultivar in the Indian communities of San Salvador, Pisac, Taray and Lamay. In the areas of Calca, Urquillos, Huayllabamba, Yucay and Urubamba, the ratio is approximately 50 percent parakay, 40 percent uwina and 10 percent other cultivars. The zone between 2,800 and 3,000 m is the maize center of the valley and probably has been since Inca times (Fig. 47). Ninety percent of the mazie grown in these two provinces is irrigated and a delay in the rainfall is not particularly crucial as it is for non-irrigated maize. On the other hand, excessive rain damages the growing crop because, with relatively low temperatures, the evaporation rate is low. The moisture causes the development of a superficial root system which then demands moisture which is unavailable at the end of the growing season. Likewise, prolonged rain during pollination is said to wash much of the pollen to the ground, preventing the wind from dispersing the pollen to the female flowers.

From Ollantaytambo (2,790 m) to the borders of the ceja (2,400 m), most of the maize grown is uwina because higher temperatures and humidity preclude the successful cultivation of parakay. Below 2,500 m uwina harvests are small; nevertheless, between 2,500 m and 2,300 m both uwina and flint and dent corn ('maíz del valle') are grown. The peasants prefer flour maize as a food but it does not thrive here because of excessive *Ustilago* rot. Flint or dent maize contain less water and are better adapted to the climate. Laurel is the most important maize of the ceja de la montaña, but below 1,900 m the chuncho cultivar is also grown. Below 2,400 m, maize is not the outstanding crop that it is in the upper valley, for diseases, pests and the absence of good flat land make root crops preferable to maize. The very high humidity promotes rot which destroys most of the flour maize that is planted. Parrots and other birds eat the kernels, as do rats, and together they destroy much maize even before it can be harvested.

Below 1,500 m, maize is more important than in the humid ceja, but it is still non-commercial. In the lower valley, unlike in Urubamba and Calca, maize is not grown on the best land. It is cultivated on the poor soils of steep slopes without irrigation and without much attention. Several cultivars of hard maize are the only kinds grown in the warm sections of the lower valley, although chullpi is sometimes grown on the coolishupper slopes. The hard maize cultivars include chuncho, laurel and pirisinko, as well as the recently introduced cuban yellow. Chunco and pirisinko show tripsacoid introgression (i.e. mixing with *Tripsacum* grass, a wild relative of maize): widely-branching tassels, limp leaves with no hairiness on the sheaths, and a canelike growth (GROBMAN et al., 1961). Maize in the upper valley has erect tassels, stiff leaves, hairy leaf sheaths and a zigzag stalk. The Quechua-speaking folk of La Convención still consider flour maize to be the only kind proper for human consumption, but they will eat flint maize if nothing else is available. Dried highland flour maize is a major trade item from the upper valley.

Maize Cultivation

Today in the Vilcanota Valley maize land is commonly prepared by oxen and wooden plow, although on steep slopes the chakitaklla is still used. Commonly the land is given a second plowing, worked at right angles to the first, 5 or 6 days before sowing. Fertilizer, mainly barnyard manure and ashes, is set in small heaps in the field before sowing, approximately 60 piles to the hectare. Although the land is best worked when soft, rarely can the land be plowed immediately after irrigation because then it is too moist.

A pair of oxen and three workers are ordinarily need to sow $\frac{1}{3}$ hectare of maize. Furrows are formed as the plow is pulled by oxen which are driven by the *yapuk*. The sower (*husk'a*), usually a woman, moves behind the plow, dropping the grains, two or three at once, into the ground. Behind the sower follows the *yakak* who puts fertilizers in the hole with the seed. Then the seeds are covered by driving the plow between the two furrows. The final step is to smooth off the field, so that the land does not dry out and so birds do not eat the seeds. Music and chicha follow the sowing.[22]

In two weeks the maize sprouts. Where germination fails, pre-soaked seeds are resown in the empty spaces. Because they are at a disadvantage in competing for light and nutrients with the earlier planted maize, these later-sown plants generally have shorter stalks and smaller ears. Maize is the most pampered crop that the peasants grow, and more work goes into its cultivation than into that of any other crop in the valley. Approximately 2 months after sowing the plants are 20-25 centimeters high and are ready for the first weeding (*hallmay*) which also includes the first *aporque*. The peasants irrigate the land 2 or 3 days before this procedure which includes several cultivational practices: a) straigthening and deepening (to about 10 centimeters) the furrows so that irrigation water flows properly and reaches the whole field; b) banking up earth around the base of the plants in the belief that maize roots are shallow and need support if the plant is not to be knocked down by winds; and c) eliminating weeds. Within 2 days after the first aporque the land is irrigated again and care is taken that the water runs through every other furrow. After this irrigation, the field is irrigated every 10 or 15 days until the rains come, usually in early December.

The second aporque (*kutipay*) follows the first by $1\frac{1}{2}$ to 2 months. As with the

22 A very well known song commonly chanted after maize sowing goes as follows (courtesy of EFRAÍN VILLENA): Kori rejawan, colki takllawan, tarpuykapuway, tarpuykapuway, chiwanway muhuta, saksa muhuta, kello muhuta, winarinanpak, wacharinanpak; (Translation: With share of gold, with plow of silver, sow me, sow me, colored seed, streaked seed, yellow seed, that it may grow, that it may yield).

first, the land is irrigated several days before the work begins and furrows are deepened and the earth is piled up around the base of the plants with the lampa. The banking promotes the development of adventitious roots in maize.

The major weeding takes place 2 to 3 months after the second aporque when the ears begin to appear on the plant. The weeds are cut with a sickle and fed to the livestock; most of the weedy brassicas by this time are too tough to serve as human food. If any bare spaces still exist due to the destruction of young maize plants, other rapidly maturing crops such as broad beans are planted.

The work of harvesting involves several discrete agricultural tasks, namely, cutting, carrying, husking, classifying, drying and shelling. Maize is cut (*sara kallkay*) beginning in the early days of May in some places, but where the crop is sown in October, harvest is delayed until early June. In any case, the peasants hope to cut the plant before the frost, for afterwards the leaves and stalks make only inferior livestock fodder. Flour maize cannot be left to dry standing uncut in the fields – a common autumnal sight in the United States – because the high water content would rot the maize. The maize stalk is cut with a sickle (*ichuna*) about 7 centimeters from the ground, working from the middle of the field toward the margins. Sometimes the maize is pulled out by the roots.

The stalks, cut or pulled, are tied into bundles (*tercios*) and carried on the backs of men or burros to a drying area (*tendal*) amid much pan flute music, chanting and drinking. If the owner lives in a village some distance away, the tendal is located behind the security of a mud wall, but if the field is near this hut, the drying area may be in the maize field itself. If the owner lives close to the field, he puts the maize in sheaves to dry (*arkuy*). The tripod arrangement helps keep rain off but occasionally the moisture in the sheaf becomes so high that the maize begins to sprout (*wiñapuska*). Thus the sheaves are turned within 15 days to avoid rot. Generally they are arranged around a watch hut made of maize stalks in which someone remains overnight to guard against theft.

The maize is husked about 30 days after being cut, mainly by women and children. The ears are first stripped from the stalks and thrown onto a large pile. A small pointed stick (*tipina*), 12 centimers long, is used to break the leafy covering of the maize ear and make stripping the husks by hand easier. As the maize is husked, it is sorted according to size and color pattern. A branched ear (*takke*) is considered to be an omen of good luck to the finder; however, the ear belongs to the owner of the maize, not to the one who husks it. Husked maize which is not to be dried (*k'ollas*) is placed aside to be used for *mote* and *sanku*. When the husking is completed, libations of *aguardiente* are poured onto the piles of maize while the saints are invoked to always bring their people a good harvest.

After being husked, the maize is laid out to dry in the drying floor (Fig. 48). A fence is built of maize stalks and leaves to protect the drying maize from animals. The drying floor is carpeted with insulating material to absorb some of the ground moisture; in Quispicanchis and Canchis, it is ichu grass, while in Calca and Urubamba it is retama branches. The maize is laid out in rows (*wachos*) according to cultivar and/or size clasisifcation. A fairly typical drying floor would have 2 rows of first class maize of parakay and uwina, 1 row of second class parakay or uwina,

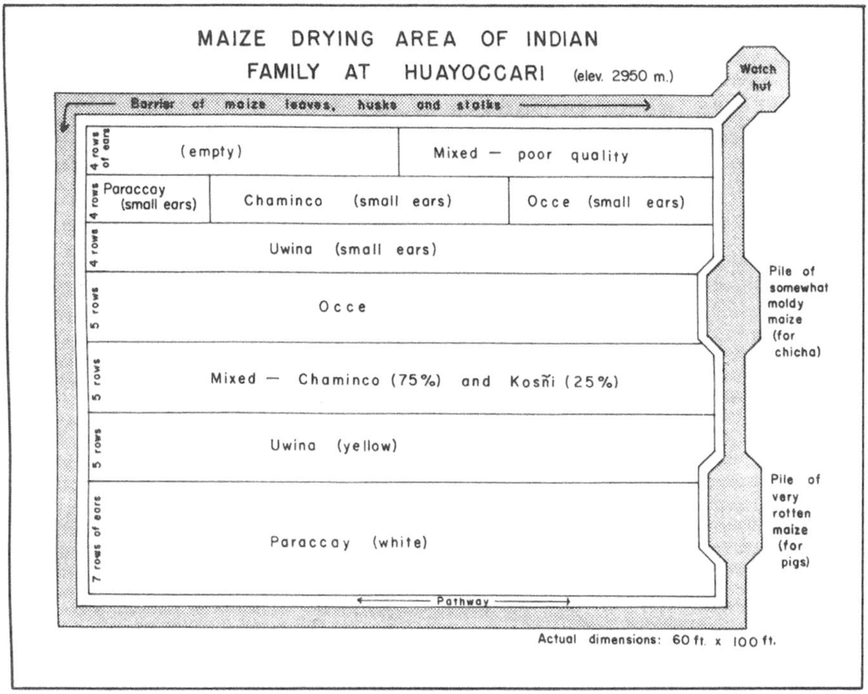

Fig. 48. Maize Drying Area Sketch.

as well as several other cultivars (Fig. 48). Flour maize has a high moisture content and requires a long period of direct sunshine to dry thoroughly; ideal are bright, sunny June days. If the weather is continuously good, the maize can be dried in 15 to 20 days, sometimes 25 days (depending on the cultivar); more often drying requires 30 to 40 days. The ears must be turned over every 2 days or so. In the lower valley, where the moisture content of the ears is relatively low, the maize is kept in the husk and hung from the rafters until needed for use.

After the maize is thoroughly dried, it is either left on the ear (*mazorca*) or the kernels are removed from the cob. Maize destined for commerce rather than home consumption is usually shelled. RAIMONDI (1898:244-5) described the shelling work (*sara saruna*) in the Department of Cuzco in the mid-nineteenth century. Ears of maize were placed on a shelling platform (about 1 meter above the ground) and trampled by Indians wearing large wooden shoes. The distance between the boards permitted the kernels to fall through and be collected on a lower shelf. These platforms are still known in the valley, but most maize, at least that which is peasant-grown, is shelled by hand.

Dried maize is stored, either on the ear or shelled. Often it is stored high in a hut on a specially built platform. In larger dwellings, a separate storeroom for maize and other agricultural products is provided.

The above discussion of maize cultivation is relevant for the main crop (*hatun tarpuy*), which is sown between about September 4 and October 15. In the Province of Urubamba, this sowing takes place in early September, but in Quispicanchis, 400 meters higher, it is in late September. In Canchis, the sowing is not until October, after the danger of frost in that area has definitely passed.

In addition to the *hatun tarpuy*, there are two other sowings of maize in the valley: a late sowing (*k'epa tarpuy*) and an early sowing (*miska tarpuy*). Farmers with unirrigated hill lands in Urubamba, Calca and Quispicanchis, often plant small parcels in maize sometime between October 10 and November 20. This late-sown maize crop depends entirely on rainfall, and if the rains are late and do not come until after Christmas, the crop fails. The peasants use a special cultivar known as *kallwa* which is both drought and frost resistant and which has a shorter growing period than the maize which is grown for the *hatun tarpuy*.

The early sowing occurs in July and early August. The distribution of *miska sara* is very limited in the valley, for although it can be grown from 2,920 m to 2,550 m, the centers of cultivation appear to be restricted to the areas around the town of Huayllabamba, Yucay and Urubamba. Above this area, frosts are a major limiting factor and although *miska sara* is sometimes planted around Calca, this part of the valley is considered to be distinctly marginal for its successful cultivation. Even within the center of cultivation, there is 20 percent risk of losing the crop since destructive August frosts frequently strike. Ample irrigation water must be available since above 2,500 m elevation, the combined rainfall of June, July and August is usually less than 6 millimeters. A little-known fungus disease attacks the *miska sara* and sometimes destroys the entire crop. Another peculiarity, not present in the *hatun tarpuy*, are the lateral ramifications which appear on the maize plants at about 4 months of age that sap energy from the plant; the farmers in the Yucay area call this *miska onkoy* (disease of early corn). Agronomists seem to know very little about the cause of these ramifications. *Miska sara* is generally harvested in January and early February. In the years when this maize escapes frosts and disease, the farmer makes a large profit because fresh ears for boiling are in great demand early in the year and they command a high price in Cuzco.

Maize Production

Maize is an important commercial product of manor estates from Hacienda Piscocucho (2,635 m) to Hacienda Ocobamba (3,450 m) in northern Canchis. Above and below this elevation, only peasants grow it and they keep most of their production for their own domestic use, and sell only small quantities in local markets. Nowhere else in the valley does commercial maize production reach such outstanding proportions as it does in the Urquillos-Huayoccari-Yucay area (about 2,960 m-2,930 m) in the heart of the 'Sacred Valley of the Incas'. Conditions of soil, climate and water combine to make this zone ideal for growing the highest quality flour maize. The soil here is loamy, of alluvial origin and with a Ph of around 7. Although the soil is naturally fertile, only with the application of organic and chemical fertilizers can high yields be maintained year after year. Within this area, production reaches, 3,990 kilograms (parakay maize) to 4,956 kilograms (uwina maize) per hectare. Outside this part of the valley, commercial maize production has lower yields and is less important. The average yields in the Ollantaytambo area and Quispicanchis Province are around 2,832 kilograms per hectare.

Hacienda production of maize in the valley follows certain well-established practices. On Hacienda Huayoccari, for example, these pratices include two aporques, eight to ten irrigations and three weedings. Plowing is now donc with a tractor, but all weeding, banking and harvesting are still done by hand labor. Insecticides are uncommon. The main organic nutrient added to the land is bird manure *(guano de isla)* imported from the coastal islands. Exceedingly high in nitrogen, guano is applied during the first aporque at the rate of 420 kilograms per hectare. Commercial maize-growing estates in the valley also require large supplies of irrigation water. Under optimum rainfall conditions, the minimum number of irrigations in the valley is five, but eight to ten are more common. Hacienda maize is always shelled after it has dried on the cob. About 20-25 percent of the seed planted does not germinate.

Much of the first class maize, which ordinarily constitutes about 70 percent of a hacienda's total corn production, is sent to other parts of Peru and is even exported to the United States. The single largest maize producer in the valley is Hacienda Huayoccari with 240 hectares of land in this region, mostly parakay cultivar. However, more typical of maize-producing estates is Hacienda Salavera at nearby Urquillos. Forty of the 200 hectares of the estate are in maize, the remainder is uncultivated hillside pasture. By using barnyard manure and imported bird manure, the owner gets a yield of 2,850 kilograms per hectare. The best land on this estate has been kept in maize production every year for the last 30 years, without change or rest. At Hacienda Compone near Ollantaytambo, 40 hectares of land are planted in maize: 27 for commercial production and 13 cultivated by the peons of the hacienda for domestic use. The commercial maize is of 2 kinds: parakay (18 hectares) and uwina (9 hectares), and both kinds are divided into first class and second class maize. First class parakay is sold to buyers from the United States, the other is sold in Cuzco.

Maize production by peasants differs from hacienda production less in technique than in yields. Most peasants cannot afford to use fertilizers, as they rarely have more than 200 soles of cash on hand and have only a limited possibility of obtaining credit. As a redult, yields are approximately 1,416 kilograms per hectare, about $\frac{1}{3}$ to $\frac{1}{2}$ the average estate production yield. A peasant uses most of the maize he produces to feed his own family. A typical disposal of his crop would be three-fourths for home consumption, one-eighth for seed and one-eight for sale.

The maize trade among peasants occurs where maize is scarce. Thus, in Canchis, where many farmers do not grow maize because of the climatic hazards, Sicuani is an important maize market for traders from Quispicanchis farther down the valley. In the lower valley, unfavorable growing conditions, pests and an ingrained preference for highland maize cultivars makes flour maize an important trade item from the Urubamba and Ollanta areas to Quillabamba. Peasant maize is sold by the individual ears, or by the *raki*, and Inca measurement consisting of a wooden box holding $1\frac{1}{2}$ arrobas (about 17 kilograms), more or less. In June and July, a raki of maize sells for 40 soles, but later, in October, when maize is much scarcer, it brings 60 soles.

Maize Food Preparation

Maize is the principal source of food and drink for at least 70 percent of the inhabitants of the valley. As far as can be determined, all the maize foods and drinks in the valley have a pre-Conquest origin. The peasants use seven methods to prepare maize: boiling, boiling and drying, parching, soaking and drying, popping, milling and brewing.

1. *Boiling* The most common method of preparation is boiling, either to make hominy (*mote*) or corn on the cob (*choklo*). Mote is kernels from which the epidermis has been removed. If the maize is young and tender, all that is necessary to loosen the epidermis is to boil the kernels in plain water, but older and drier maize must be boiled in a mixture of molle (*Schinus molle*) ashes and water to loosen this covering. In either case, the kernels are then rubbed in the palms of the hand to remove the epidermis and later this *maíz pelado* is air dried. Young, unripe maize is also picked to be boiled and eaten corn-on-the-cob fashion; the miska harvest is destined for sale as green corn and is served prepared in this manner.

2. *Boiling and Drying* Very ripe, fresh maize ears, boiled in water and dried in the sun are known as *chochoka*. When milled, chochoka is made into tamales, mush and added to soups.

3. *Parching* Maize that is parched in a clay vessel over a fire is known as *kancha*. The uniform heat turns the interior of the kernel white and makes the outer covering crunchy and delicious. Toasted maize is a finger food and is frequently taken on journeys.

4. *Soaking and Drying* This process not only imparts a special flavor and texture to maize, but it also is a means of keeping the grains soft for a long period of time. The husked maize is placed between two layers of maize leaves in an irrigation ditch or stream. Stones are placed on top to prevent the current from carrying the ears away. After being soaked in water for 30 days, the maize is removed, dried in the sun and milled into a mush. It is known as *tokosh* and is a rare method of preparing maize in the valley today.

5. *Popping* Popcorn (*maná*) is made for special occasions such as fiestas, but it is not a common form of maize preparation. Around Lake Titicaca the corn is popped with hot sand, but in the valley the kernels are popped in clay vessels heated over a low fire.

6. *Milling* Dried maize may be either rough ground to make *lawa* (*mazamorra* in Spanish), or it may be finely ground to make mush (*api*). Wet maize is never ground. If the dried maize has first been parched, the flour is known as *allpa*. Milled maize may be consumed as is with water or milk, or it may be used to make more elaborate dishes. One such special dish is *sanku* which is prepared by toasting

finely-ground maize, mixing it with molasses, and forming it into balls. Today these sweetish hunks are traditionally eaten after the maize harvest. GARCILASO DE LA VEGA (1960:269) relates that in Inca times sanku was the main dish at the solemn banquet for the festival of the Sun (*Inti Raymi*). This pattern of use suggests that this food has a eucharistic significance at such ceremonial times. Finely milled maize made from chochoka, as explained above, is used to make the dough for tamales. Unlike in Mexico, tamales are a special food in Peru and are made only in April and May. Maize tortillas – so basic a food in Mexico – are not made in Highland Peru.

7. *Brewing* Maize chicha remains the common beverage for the indigenous population although the more acculturated cholos drink bottled barley beer in place of chicha. Most flour maize, except parakay, can be used for chicha, but chaminko, kulli and uwina are the preferred varieties. The first step in the preparation of this yellowish-brown beverage is to sprout the maize kernels; this is accomplished by soaking the maize in water for several days and then placing it in a pot or small stone-lined pit (*kocha*) covered with moist maize leaves or straw. The maize remains there 10 to 12 days, sometimes less, until it has sprouted (*jora* in Spanish and *wiñapo* in Quechua), providing enzymes to change starches to sugar. Formerly unsprouted maize was masticated (COBO, I:162-3; TSCHUDI, 1847:261), the saliva supplying the needed enzymes. MARCOY (1875:130) commented on masticated corn in his nineteenth-century travels to Combapata, but inquiry in that locality indicates that this practice stopped around the turn of the century.

The sprouted maize is dried in the sun for a day to prevent the germination from proceeding farther. It is then ground, either by a saddlestone if for home preparation, or by gasoline-powered *jora* mills in the larger towns if large quantities of chicha are made. Ground *jora* is not a common trade item, perhaps because of the ease with which it can be adulterated. Chicha making proper begins when the jora flour is mixed with water to form a dough (*peka*). Boiling water is then poured slowly onto the dough, though in some chicherías, jora flour is simply added to the boiling water. In either case, after 2 hours of boiling and stirring, the liquid is poured into large wide-mouthed pots where it remains overnight. Many consider this liquid, known as *punuy aka*, to be delicious and drink it as is. More typically, however, the punuy aka is boiled again and mixed with water. Eventually it forms a thick substance (*hanchi*) which is then strained through a straw basket or a piece of cheese cloth. The thin liquid which passes through is known as *upi*; the dross (*sutuche*) which remains in the basket is fed to guinea pigs or saved to be used as yeast in breadmaking. The *upi* is then poured into a large but narrow-mouthed jar (*urpo*) where it ferments; in 3 days it is ready to drink. Chicha becomes bitter if not drunk within a day or two; unlike some alcoholic beverages, it does not improve with age.

Besides the common, everyday chicha, several variations are made in different places and at different times. *Sutuy aka* is chicha made with both maize and barley, and is prepared particularly for fiestas in northern Canchis. The beverage ferments a month and sugar is added to make it very sweet and thick. *Pampay aka* is simply chicha that has matured underground for 6 months with the top of the narrow-mouthed jar closed with pieces of dough. *Tekti ruway* is a milky-looking non-alcoholic chicha made with maize seed (not sprouts), sugar, quinoa and other products such as peach stones. *Frutillada* is made in Urubamba and Yucay during the strawberry harvest in December and January; it is chicha mixed with strawberry juice and a little rum.

By no means do all peasant households make chicha. In the maize region, there is more of a tendency to make it for home consumption, but even in these villages are women who specialize in brewing and selling this muddy-looking drink to the public. The common way to announce a chichería (place where the beverage is sold) in this non-literate society is to attach bright flowers to the end of a long pole: a bunch of all red flowers signifies that the chicha seller is married, white flowers indicate that she is single. In the absence of flowers, the panache may be pampas grass, maize leaves or a cloth rag. Urubamba is the largest chicha-making center in the valley, and most of the chicha vendors at the big August fair at Tiobamba and the September fair at Huanta are *urubambinas*. In areas above 3,400 m, more sugar cane alcohol is consumed which, of course, has a much higher alcoholic content. In the lower valley, below 1,500 m, chicha making also becomes relatively less important because the local maize cultivars do not make satisfactory chicha, and because bottled barley beer has become widely accepted among the peasant class.

Other Food and Non-Food Uses of Maize

Beginning in February, the stalks of maize are chewed for their sugary juice. Known as *wiro*, the stalk is strictly a by-product of maize cultivation and are so used only after the maize has been harvested for the ears.

A parasitic growth on maize, known as *hatupa* or *pacorma*, is used as human food. This edible spongy mass, occurring as a tumor on the maize ear or stalk, is caused by corn smut (*Ustilago maydis*). As the fungus matures into the spore-forming stage, it begins to turn black. Most Indians relish its delicate flavor in stews and most cholos also consume it without compunction. Mestizos, however, disdain it because of its cultural associations. Those who eat hatupa voice no fears about any adverse effects. Some hatupa-infested maize is fed to cattle and pigs.[23]

The maize plant has a number of uses to man in the valley besides the grain it yields. Foremost among these is the use of the dried stalk together with the dried leaves, all known as *chala*, as a livestock fodder. Some maize chala was undoubtedly fed to llamas and guinea pigs in Inca times, and although today some maize on the Coast is grown exclusively for its chala, in the valley the fodder provided has never been as or more important than the grain. Widespread use of chala for stock feed came after the Conquest when the Spaniards introduced horses, burros, mules, cattle and other animals. After the maize has ripened, the whole plant is cut

[23] Human excrement is the single most important source of food for pigs and therefore these domestic animals play a big role not only in the energy cycle but in sanitation as well.

with a sickle and carried by men to the patio of the dwelling. It is first used to make barriers surrounding the drying maize. Later, the chala is stored on platforms (*pirwas*) to keep it out of animals' reach and to save space.

Corn husks (*panka*) are used as fuel, for wrapping foods and to fill mattresses. Corn cobs (*koronta*) are used as fuel. Corn silks (*sarachukcha*) have medicinal value as a remedy for bladder trouble.

Andropogon citratus DC., *hierba luisa* or lemon grass, is an aromatic, lemon-scented grass that is found below 1,500 m in the valley. In hilly coffee plantings it is grown to stabilize the soil. Lemon grass volunteers readily. It is used in folk medicine as an unspecific home remedy.

Sugar Cane

Saccharum officinarum L., *caña de azucar* or sugar cane, may have entered the lower Vilcanota Valley before the beginning of the seventeenth century. CIEZA DE LEON (1883:231), travelling in 1551, did not mention sugar came in referring to the warm valleys of Cuzco, although he wrote of such Old World plants as the orange, lime, fig, the vine and banana. By 1609, however, sugar cane had already become important in the eastern valleys of Cuzco; in that year COBO (I:405) visited Cuzco and noted that the best sugar in Peru was made in the warm valleys of that area. In 1610 BALTASAR DE OCAMPO (MAURTUA, VIII:267) wrote of the *ingenios* or sugar mills in the Vilcanota Valley which were worked with Negro labor. A century later sugar cane had spread to the Machiguenga Indians (QUEVEDO, 1900:499).

The lower valley became famous for its sugar cane, particularly after the Jesuits organized agricultural enterprises, notably Hacienda Santa Ana near present-day Quillabamba, connected with their missionizing work. A seventeenth-century Jesuit manuscript (ANONYMOUS M.S: n.d.) cited the advantage of not needing to replant cane every year. Disadvantages were long distance to market and the shortage of barrels to send the annual harvest of 35-40,000 kilograms of sugar to Cuzco. The Jesuits were expelled from the valley in 1767 but the owners who took over their land in the area maintained the agronomic emphasis on sugar cane (LAVANDAIS, 1851:1022).

Cane cultivation declined greatly during the turmoil of the movement for Latin American independence from Spain. But by 1830, in increasing demand for sugar cane alcohol revived cane production because of the introduction of cheaper milling processes and the generalization of alcohol consumption, previously restricted by the Spanish colonial authorities. In the early nineteenth century hydraulic sugar presses having iron or bronze crushing cylinders replaced most of the old mills which were turned by oxen and which used trunks of the *wilka* tree for cylinders; the sharp protuberances of this trunk facilitated crushing the cane. Although oxen were still used on some cane estates in the lower valley well into the twentieth century (HARDY, 1914:510), the much more efficient hydraulic mills produced alcohol cheaply enough so that even the poor Sierra Indians could purchase it, and it could be economically exported from the lower Vilcanota Valley. This increased the importance of sugar cane cultivation. When ANTONIO RAIMONDI (1898:48) travelled in the lower valley in 1856, he noted that sugar cane was the dominant commercial crop at the Haciendas Echarati, Media Luna, Maranura, Chinche, Chaullay, Santa Ana, Ccolpani and Huadquiña. In short, by that time most of the haciendas in the valley depended largely on cane for their cash crop.

Varieties

The first sugar cane cultivar taken into the Vilcanota Valley was that known as *criolla* or *serrana* or (by the peasants) *suave*. Originating in India, the Crusaders carried this variety of cane to the Middle East and thence to Southern Europe. The Spaniards grew it in Andalusia and then in the Canaries. From the Canary Islands, it was brought to the New World. Criolla cane is short, 1-1$\frac{1}{2}$ meters tall, with a soft yellow stalk, rich in sucrose. But today, because of its shortness and low yields, it no longer has any commercial value in the valley. It is grown only by the Indians who prefer it because it is very sweet and soft, therefore easy to press. This latter characteristic is important as the peasants here have no sugar mills and hence extract the juice by very primitive methods. The Indians chew the raw cane itself as a confection (*miski wiro*), sucking the juice and spitting out the pulp. Also they prepare cane chicha (*cuñazo*) by boiling the cane juice and allowing it to slightly ferment.

The most important variety for more than 100 years in the valley has been *caña india* or *borbón*. Originally from Tahiti (DEERR, 1928:22), this cultivar was taken by the French in 1768 to Mauritius and Bourbon (today Réunion) from whence it got its general name. Between 1810 and 1840, it was taken to Peru where it rapidly replaced criolla as the commercial cultivar. Borbón is at least $\frac{1}{2}$ meter taller than criolla and has lighter-colored leaves, but its sucrose content is lower and it does not yield as long. RAIMONDI (1898:48), travelling through the lower Vilcanota Valley in 1856-7, observed that 'caña de la India' had replaced criolla on the big Hacienda Huadquiña even though criolla was sweeter and softer. Borbón cane, however, had the great advantage of higher yields and a shorter maturing period (14 months). Yet at the same time it is shallow-rooted and hence not a drought-resistant cane, and it is also very susceptible to fungus disease (DEERR 1928:25).

Beginning about 1850 hardy canes of the Cheribon group of Javanese origin, known locally as *cristalina* cane, were brought to Peru from Louisiana. In 1856 RAIMONDI 1898:31) noted the presence of a mutant of this cane at Chaullay where it was known as *kulli* or *morada*. It is now very rare. Another mutant of the same variety grown occasionally is *rayada* (international name: striped preanger) which was introduced from Louisiana in 1907 (GONZÁLEZ TAFUR, 1928:12). This cane is still grown in the upper reaches of some of the lateral valleys of the Vilcanota bacause it is precocious and resistant to cold. In the valley proper, both of these mutants are grown only by the peasants around their chacras for their own domestic requirements.

Since the 1950's, several cane growers in the lower Vilcanota Valley have tried the P.O.J. (*piojota*) series of varieties originally from Java, but these 'modern' canes are hard and not easily pressed in the antiquated hydraulic mills that are still used on most cane haciendas; for this reason, the softer borbón is still preferred although its lower yields are acknowledged.

The upper limit of commercial sugar cane cultivation in the valley is 1,550 m. Above this elevation, up to 2,100 m, sugar cane is grown on peasant chacras for domestic purposes. At 1,000 m cane matures in 1 year; at 1,300 m it is ready in 18 months and at 1,500 m it requires 18 to 24 months to mature. Below about 1,250 m cane begins to flower when mature which causes the sucrose of the stem to be transferred to the seeds. To prevent further loss of sucrose, the cane must be harvested immediately and as rapidly as possible. Cane grown above the zone for flowering thus has a higher sucrose content which offsets the longer growing period needed at the higher elevation.

Cane Cultivation

In most places in the valley cane cultivation still uses much human and animal labor. The old method of preparing the land by men wielding the *lampa* (or sometimes pick) has given way to plowing with oxen. Most haciendas have replaced wooden plows by steel plows, but the work of planting still is performed by peons who put multinodal pieces of cane in the furrows. During its growth cane is weeded and banked by hand two or three times a year. The first cutting, done when the plants are about 3 meters high, is referred to as the *planta*; the roots are left to sprout again and the second cutting, called *soca*, ripens more quickly than the first and grows a meter or so shorter. The third cutting (*resoca*) has a smaller yield than the first two. The same cane rootstock may be kept as long as 10 straight years before being planted again, but if plentiful labor is available, the field may be burned and new rootstocks planted every 3 years.

All commercial cane in the valley is irrigated from May to October. The common practice is to grow sugar cane on the same lands for many years without crop rotation or fallowing; for this reason, cane yields here are one half to one third those of the coastal haciendas. The scarcity of labor has caused a great decline in cane plantings in the valley, and the crop may eventually disappear as agrarian reform changes the land holding pattern.

Unlike in certain other cane-growing areas of Latin America, such as Colombia or Guatemala, sugar cane in the Vilcanota Valley never has been a serious component of peasant agriculture. This cannot be explained by any low relative usefulness of this plant, for it yields food, drink, fodder and fuel. Rather, the peasants here have none of the irrigated flat land which is necessary to successfully grow cane, and arrendire contracts in the past specifically forbad the tenants to grow it, because of the competition in periodic labor requirements (HOBSBAWM, 1969:38). Where peasants do cultivate cane, it is a question of a few plants near the hut, with the sweet juice pressed out by primitive devices.

Use of Sugar Cane

During the seventeenth, eighteenth and part of the nineteenth centuries, the main products made from the cane grown on haciendas in the Vilcanota Valley were crystalline sugar (*azúcar rubio*) and crude brown sugar (*chancaca*). But today these products have been replaced by rum (*aguardiente*) into which virtually all the cane is now made. Crystalline sugar making had ceased by 1920, because the low quality and low value sugar that was produced on the estates and carried by mules to the market at Cuzco could not compete with the refined sugar brought by railroad from plantations in the Arequipa region. In 1916 (BUES, 1917:17) only about 90,000 kilograms of crystalline sugar, half of which came from Hacienda Echarati, were sent out from La Convención, whereas over ten times that amount of rum was sent out. Chancaca is made by boiling sugar cane juice to clarify it and to evaporate the water content; the syrup is then cooled in special molds so that cakes of solid, unrefined dark brown sugar are formed. Chancaca was more important than even crystalline sugar in colonial days, and, as recently as 25 years ago, is was still the most common sweet eaten throughout the Sierra. But subsequently its production declined greatly and in the larger towns commercial candy has replaced it. Indians now eat chancaca for energy on arduous journeys. Unlike small farmers in Guatemala or Colombia, those in Southern Peru do not make a distilled liquor from this unrefined sugar.

Rum is now the most profitable and thus, the most common, product of sugar cane. Despite government taxes on alcohol production – taxes which many haciendas manage to partly circumvent – an estate usually can earn over US$200.00 per hectare of cane grown. With profits like this, almost all the cane grown in the valley is turned into rum, and the two have become virtually synonomous; estate managers refer to yields of sugar cane in terms of quantities of full strength aguardiente per topo.

To make rum, juice is pressed out of the cut cane in the mill. The bagasse is used as fuel and as livestock fodder. The juice (*caldo*) is run through a system of tubes to large copper tanks (*pailas*) where the 'cooking' takes place. The boiled liquid is diluted and left to cool, then passed through other tubes to wooden casks (*pipas*) where it ferments 4 to 5 days, becoming 'chicha'. If the cane was not completely ripe when cut, fermentation takes longer. From the pipa, the liquid is moved to the *falca*, a kind of still in which the 'chicha' is converted into rum. The alcohol is made at 97 percent, but watered down to 50 percent to meet governmental regulations.

The rum is shipped from the Vilcanota Valley to Cuzco by rail or truck, though prior to 1930 it had to be carried out on muleback. From Cuzco it is distributed throughout the Departments of Cuzco and Puno. Cuzco Department also receives some alcohol from the Abancay district (Dept. of Apurímac), while Puno Department receives part of its supply from Arequipa. Indians and cholos consume rum in large quantities, and even those living in the most isolated places receive it at least for the local annual fiesta which invariably turns into a drunken orgy. A milder beverage made non-commercially by peons is *cambrai*, sugar cane juice that has been boiled and fermented only.

Pennisetum clandestinum Hoechst, *kikuyo,* is a tropical grass originally from East Africa which agronomists say was introduced into the Vilcanota Valley some time after 1950 as a planted pasture species. Since then, it has aggressively spread into crop fields and along roadsides and irrigation ditches in the entire valley below 3,995 m, taking over almost all pasture land and invading about half of the unirrigated land. Reproduction in the kikuyo grass is asexual in the upper valley for the panicle inflorescence does not develop here, and its spread is due to its vigorous subterranean rhizome and to the adventitious roots which appear at each node. Although the surface part of the plant is killed during the low-sun period in the upper valley, the subterranean rhizome survives. Kikuyo is perhaps the greatest plant pest in the valley today and much time and effort is expended in its extirpation. Arable land having a cover of this grass is very difficult to plow, and in grain fields it impedes the development of the superficial roots of cereal crops. For the Indians, however, it is not a totally noxious pest because it does supply fairly good forage for their animals.

Paspalum gayanum L., *grama*, forms a short but dense turf in poorly drained areas in the upper valley and is eaten by cattle.

Lolium tenum L., *cerisuela* is a noxious weed in wheat and barley fields. The seeds are often mixed with grain inadvertently, which is undesirable because they are poisonous in large amounts, narcotic in small quantities.

Bromus Haenkianus, cebadilla, is a wild grass found below 3,500 m which is used as a livestock fodder.

Festuca dolicophylla Presl., *chillwa*, is a perennial bunch grass which is the outstanding pasturage at elevations above 3,700 m. Althought its food value may not be high, it is regarded as one of the best native forage plants because it can support animals in the dry season due to its drought and frost resistance. It is also used as thatch.

Festuca dichoclada Pilger, *sailla*, is a perennial grass used as forage and thatch in the upper valley.

Poa pratensis L., *gramilla* or Kentucky blue grass, is an aggressive perennial grass from Eurasia which tends to invade fields below 3,600 m. It is used as a livestock fodder.

Avena sativa L., *avena* or oats is a cultivated European grain that was not greatly accepted into the folk agriculture of the valley, even though yields in test plots prove that the valley has ecological conditions favorable for it. Near the limits of agriculture some estate owners raise forage oats (*avena forrajera*) to feed to their horses. Occasionally peasants cultivate oats on land ordinarily devoted to wheat or barley. The oats are dried on racks near their huts and the grain is used in soups and the stalks used as fodder. Perhaps because they are difficult to prepare for human consumption oats are only infrequently grown today in the valley, although since 1964, more land is devoted to this crop. The only oats that many people in Southern Peru know are American or Canadian rolled oats which are sold in tins in some of the larger towns, and 'Quaker' has become the nominal term for oats.

Avena barbata L., *avena silvestre* or wild oats, is a common weed in wheat and barley fields which is collected and fed to livestock.

Secale cereale L., *centeno* or rye, is almost unknown by the peasants in this part of Peru. It was introduced (COBO: II:417) in the sixteenth century, but there were better crops and rye apparently was not needed. More important, however, is the fact that rye is a northern European crop and was never part of the southern Spaniard's diet. Had the Spaniards themselves valued rye as they did, say, wheat, it undoubtedly would have gained acceptance. Certainly rye thrives on the rocky soils and low temperatures of the upper valley as shown in experimental plots. It would, however, compete with quinoa and potatoes for these sites and thus would have no ecological niche of its own.

Barley

Hordeum spp., *cebada* or barley, is the other small grain besides wheat which has been so important in the valley since the sixteenth century. Barley had limited mention in the historical sources, although it may have been introduced earlier than wheat. Pizarro landed at Tumbes in 1532 and may have brought barley as feed for his horses. An any rate, barley was undoubtedly an important commercial crop during colonial times for horse and mule feed. In high altitudes where the grain did not mature, for example, on the road to Potosi, COBO (I:409) tells us that barley was grown only for fodder. In the area around La Raya Pass (above 4,000 m) barley may have been important in those days for, by edict of the Viceroy Toledo (LEVILLIER, 1925:VIII:130), barley was required to be grown at tambos. Today barley is more important to the peasants for human food than for animal feed.

Cultivation

Barley is commonly sown after wheat has been planted but they are harvested about the same time. In Canchis, November is the barley sowing month and in Tinta barley is sown on or after the Festival of San Martín, November 11. In Quispicanchis, Calca and Urubamba, it is sown in December. Some 'winter' barley, sown in May right after the maize or potato harvest, is grown for fodder to be cut in October. Although the people prefer to eat wheat, in many places barley is favored as a crop because it is resistant to *Puccinia* disease and it does better on the salty soils of some valley flats. In Canchis, the valley slopes are sown in barley; in Quispicanchis, the slopes are sown in both barley and wheat; in Calca and Urubamba, they are mostly in wheat. Barley has a greater cold tolerance than wheat and this basically explains the above distribution. Barley is also generally more disease resistant than wheat but it is susceptible to several species of smut (*Ustilago*) and to a *Helminthosporium* fungus.

By far the most important kind of barley grown in the Vilcanota Valley is a 6-rowed cultivar known as *común* or *corriente*, belonging to the species *Hordeum vulgare*. Común barley possibly was brought to Peru from Spain in the early colonial period. It is an awned variety, has large kernels, a thick hull and is relatively rich in protein and relatively poor in starch. When perfectly ripe, común barley loses its grain easily so it is harvested before it is completely mature. This plant is very cold resistant and the Indians grow it higher than any other Old World food crop (3,910 m south of Occobamba). It is also drought resistant and therefore seldom irrigated. Commonly it is seen growing on the steep slopes of the valley in an attractive checkerboard pattern.

Particularly in Canchis, común barley forms an important part of the native diet. It is most commonly toasted and coarsely ground to make *chakepa*, which is added to a broth to make a soup, or it is an ingredient in stew with brassica greens and potatoes. Toasted barley is also ground into a fine flour used to make mush or small flat loaves of bread. Bread which is sold commercially in the valley often contains as much as 30 percent barley flour mixed with the wheat flour. Chicha made from barley is commonly prepared in Canchis by a process similar to that for preparing maize chicha. The *chicha de cebada* of Combapata has become well known throughout the area and is made in most households. MIDDENDORF (1895:III:455), travelling through Checcacupe is the late nineteenth century, noted that their chicha was made of two parts of maize malt and three parts of barley malt. A reference (GONZÁLEZ 1833:51) to *aguardiente de cebada* suggests that whiskey was made at Urubamba in 1833, but the low cost of sugar cane alcohol stopped such attempts at hard liquor production from barley.

To some extent barley is grown as an animal fodder. Land from which potatoes have been harvested is often sown with barley which grows, irrigated, through the dry season to be harvested green in October or later. Green barley (known as *alcacer* or, by the Indians, *kacho*) is sold in small bundles for livestock and guinea pigs. Guinea pigs are kept in every hut and can derive all their moisture needs from the green forage fed to them (GADE, 1967: 221). Besides green barley the dried barley straw is also fed to livestock after the regular harvest. Barley straw is softer and more nutritious than wheat straw, although both are said by agronomists to be poor in digestible proteins.

K'alauma, another cultivar of *H.vulgare*, is grown to some extent in the Vilcanota Valley above 2,800 m. Unlike común barley, the lemma and palea of *k'alauma* do not adhere to the caryopsis, a 'naked' condition called by the Quechua equivalent of 'bald.' This plant is most easily recognized by the hoods which are attached to the lemma by a short awn-like appendage and by its long, wide, drooping leaves. HARLAN (1918:7) believed that the hooded varieties of barley originated in the Nepal barley, first noticed by Europeans in 1839. When and from where this barley came to the Vilcanota Valley is not known, but it probably was less than 100 years ago. K'alauma barley has considerable importance above the rim of the valley from Urquillos, and it is possible that it was from this point that is was subsequently dispersed to the Vilcanota. In the valley it is now uncommonly grown on hillsides, rarely in the valley bottoms. K'alauma is cold tolerant but not hail resistant, and perhaps for this reason it is less cultivated in Canchis than one might expect. However, the Canchis Indian community of Queromarca grows considerable quantities of this kind of barley. It threshes free like ordinary wheat, and the grain, rich in starch, is eaten toasted and triturated. Some barley bread is also made from it. The straw is highly valued for forage because of the relatively large leaves.

Hordeum distichum L. is a 2-rowed barley in which only the central spikelets are fertile; the two lateral spikelets atrophy. Two-rowed barley has increased in importance in the valley particularly in the past 20 years because many cholos and most mestizos in the Deparment of Cuzco now drink bottled beer rather than maize chicha. *Chevalier*, a cultivar imported in improved form from Europe in the mid-1920's has been grown as *cebada alemana* to be sold to the German-owned brewery in Cuzco which provides, without interest, the seed and fertilizer under contract to the farmer. Malt made from 2-rowed barley is preferred to that from 6-rowed barley because the latter releases an excess of certain protein substances during brewing which makes the liquid cloudy.

Two-rowed barley is grown from about 2,515 to 3,400 m. The zone with the highest yields and largest hectarage is between Pisac and Chilca, centering on the Ollantaytambo area. Here beer barley is grown under irrigation on the best soils, those usually reserved for maize. Occasionally the yields in Ollanta are as high as 3,000 kilograms per hectare, but more commonly they are only a third of that. Yields decrease below 2,700 m and above 3,000 m, and above 3,500 m the malting properties of barley grown at this elevation are inferior.

The seed of this 2-rowed barley is sown broadcast. The amount of seed sown varies from 100 kilograms to 360 kilograms per hectare; more seed per unit area means less weeding. Small peasant farmers have better yields than estate producers because they sow thicker and weed the fields more carefully. Smut (*carbón*), which occurs with excessive humidity, is the only disease which presents much difficulty; *Puccinia* does not attack this barley. The amount of land in 2-rowed barley varies from year to year depending on the price of other crops; if the price of maize is high, less beer barley will be grown. Considered by decades, however, the cultivation of 2-rowed barley has increased greatly according to agronomists. This may be explained not only by the brewery's aggressive propaganda, but also by the widespread occurrence of *Puccinia* disease in wheat and by the desire of an increasing number of peasants to participate in a cash economy.

To a small extent, Indians grow 2-rowed barley for their own food needs, although most still prefer the comûn barley for this purpose. In recent years, however, the brewery in Cuzco has engaged in a campaign to eliminate comûn barley and thus to prevent the otherwise inevitable mixing of comûn barley with the 2-rowed material sold to the brewery by giving small quantities of the 2-rowed barley to the Indians. In 1953 the brewery estimated that 2-rowed barley constituted 4 percent of the barley cultivated in the Department of Cuzco; in 1963, the estimate had risen to 16 percent. (Annual Report, Compañía Cervecera del Sur, 1962-63). At this rate, comûn barley will be very rare in the valley within 25 years.

Wheat

Triticum spp., *trigo* or wheat, was introduced by the Spaniards to satisfy their taste for wheat bread. A document dating from 1493 (RODRIGUEZ DEL RIVERO, 1947:173-5) indicates that Columbus carried wheat to Hispaniola from Cádiz; it is likely, however, that the Peruvian introduction came directly from Spain some 40 years later. Both GARCILASO DE LA VEGA (1960:442) and COBO (1:406) relate the romantic but questionable account of how wheat, once it reached Peru, was carefully grown in small pots by a Spanish lady who distributed a few seeds to her neighbors until, gradually, it was planted in large fields. Although wheat was in Cuzco by 1539, there was still no wheat bread and the Spaniards continued to eat maize cakes (GARCILASO DE LA VEGA, 1960:442). But by 1551 wheat was an important crop: an injunction by the Municipality of Cuzco forbade the reselling of wheat for profit motives; another order demanded that since the price of wheat was low, the breadmakers should sell their bread at accordingly low prices (COVARRUBIAS POZO 1961:320). The first bigtime profiteers in wheat appeared in Cuzco in 1586 (MONTESINOS, 1906:101), forcing the municipality to fix staple food prices.

The outstanding importance of wheat in the early years of Spanish occupation was due to the Spaniards' demand for it and, later, the mestizos demand. The latter people readily adopted wheat as a food since bread enjoyed the prestige of being the 'food of the Conquerers' and in this way acquired a cachet of superiority. Later, in the seventeenth century, Indians grew wheat only because it was among the products with which tributes were paid. The religious orders owned much land in the valley and rent was paid in kind by the tenants. It is difficult to say when the Indians started growing wheat for their own use, perhaps not until the eighteenth century (JIMÉNEZ DE LA ESPADA, 1956:174).

Present Cultivation

Today wheat is grown in the Vilcanota Valley between 2,800 and 3,800 m. Other things being equal, the wheat grown at lower elevations is believed to contain more gluten than that grown at the upper altitudinal limits. The upper limit, which has moved up valley at least 500 meters in the past 30 years due to the introduction of a new cultivar, is set by killing frosts in all months. The highest wheat field is at 3,905 m above sea level, very close to the upper limit of barley cultivation. Actually wheat is not well adapted to the prevailing climate of any part of the valley. The kernels form and ripen best when temperatures are high (between 25° and 35° C.) and temperatures never reach this level in the area in which wheat is grown. Furthermore, the lack of hot, dry summers is said to cause a low gluten content in all kinds of wheat because under lower temperatures, starch accumulates faster than proteins. The general tendency is for a hard wheat to produce kernels of a softer type after having been grown in the Sierra for several years. The coastal miller or baker does not think highly of Sierra-grown wheat. The abundant presence of wheat rust is a major hazard to wheat growing in the Vilcanota Valley, and the high occurrence of this disease caused by high humidity, prevents wheat cultivation below 2,800 m.

Generally known by the peasants as *roya* the Indians also call wheat rust *uchucuta* because of its fanciful similarity in appearance to ground dried capsicum pepper (uchu=capsicum pepper). Barberry (*Berberis* sp.) is the host plant of wheat rust and is common in the upper valley along roadsides and on top of adobe walls where it is planted as a hedge. The peasants generally do not understand the connection between barberry and wheat rust, and therefore they see no reason to eliminate barberry. Wheat rust has been known in Peru since at least 1746 (GARCÍA 1936:36). In the Vilcanota Valley there are at least two species: *Puccinia glumarum (roya amarilla)*; and *P. graminis tritici (roya negra)*. Roya amarilla has apparently existed in the Department of Cuzco for a long time but only in 1928 was it scientifically identified (GARCÍA 1936:43). It is more common and devastating than roya negra. Wheat rust of any species not only reduces the number and weight of grains per spike, but also lowers the quality of the flour. An ordinary attack of *Puccinia* can reduce yields by one-third. In the valley it begins to appear in January and February during the height of the rainy season. In 1932 and 1933 wheat rust destroyed most of the wheat planted in the valley; this spectacular decrease in yields is seen in Table 2 below. This agricultural disaster motivated the Ministry of Agriculture and other agencies to introduce new rust-resistant cultivars of wheat, including *marquis* from North America and *mentana* from Italy. These gained wide acceptance among both commercially oriented hacendados and subsistence-oriented peasants, and it is reasonable to assume that certain old cultivars of wheat became rare and perhaps were even extinguished during this time and that barley plantings were greatly increased.

Table 2. Wheat production for Selected Years in the Valley in M. Tons.

	Urubamba	Calca	Quispicanchis	Canchis
1929	2,363	908	2,301	3,349
1930	621	897	1,597	13,357
1931	1,138	218	2,017	10,037
1932	27	80	21	857
1933	82	280	215	248

(source: GARCÍA, 1936:36)

Kinds of Wheat

Almost all the wheat in the Vilcanota Valley is one of two species: *Triticum durum* and *T. sativum*. Of the more primitive wheats, spelt (*T. spelta*), known here as *capli* is cultivated sparingly in some of the more isolated lateral valleys back from the main Vilcanota depression. Eikhorn, emmer, poulard and club wheats appear to be absent from the valley, although they are still found in Spain (GADEA 1954) where the wheat flora is much larger than in Peru. It seems possible that even sixteenth century Spaniards did not have high regard for these lesser species, of wheat and thus did not make much effort to establish them in Peru.

T. durum is sometimes called *chumpi* (brown) wheat by the Indians because of its brown-colored kernels. This wheat is characteristically long-awned with solid internodes and it tillers very little. Chumpi wheat is tall and therefore provides more fodder that wheat with short stalks, one of the several reasons why Indians grow this kind. This wheat is conspicuously low in gluten quality and content, and makes flat, hard breads. Abundant, high-quality gluten permits the dough to retain the gases which form inside, giving a raised, soft bread. The durum wheat commonly grown in the Vilcanota Valley is known as *barba negra* (or *yana barba*) by the farmers. This tetraploid variety is cultivated from 2,800 m to 3,500 m, both on hillsides and under irrigation. It is identified by extremely long barbed awns, measuring 25 centimeters, and dark-colored glumes which are covered with a light, gray-colored pubescence. The Indians believe that the long awns discourage birds from eating the grain; on the other hand, the awns have sharp barbs which can injure the stomachs of livestock. The light brown caryopsis is elliptical-ovoid and about 8 millimeters long.

The Indians value barba negra because it has a soft grain which can easily be ground with saddlestones and with millstones moved by water wheels (GADE, 1971). It is also considered to be drought-resistant, an important onsideration for any unirrigated agriculture, especially on dry, steep hillsides. The vegetative cycle is long, in some areas as long as $7\frac{1}{2}$ months, and many kernels are plasmolyzed ('granos chuzos') because of low temperatures, mainly in May and June. The yields are low, typically between 600 and 700 kilograms per hectare. Although durum wheats were scientifically distinguished from other wheats only near the end of the eighteenth century, this wheat may have been introduced into the valley in the sixteenth century.[24] Barba negra would have been eliminated from valley agriculture long ago if it were not for its soft kernels which can be easily ground with a primitive grinding technology.

Another chumpi cultivar in the valley is *candeal* (also known as *candaño*) which may have been brought to the valley in the sixteenth century from Spain, although this would be difficult to prove. Like barba negra, its long awns, drought resistance, and soft kernels offset its disadvantages of a long vegetative cycle and prevalence of 'chuzo' grains, low yields (650-700 kg/ha) and susceptibility to wheat rust. Candeal has an ecological niche in the valley from around 3,500 m to 3,750 m in southern Canchis.

Most of the wheat in the valley today includes cultivars of *Triticum sativum*, the common bread wheat. The antique *sativum* cultivar in the valley would appear to be *barba blanca* which is grown down to about 2,800 m on dry hillsides. Much more important are the various hybrid cultivars of this species which date from the 1930's when *mentana* (also known as *frontana*) was introduced from Italy. This hybrid winter wheat (in Italy) is resistant to cold, although less resistant to drought than candeal. It is awned with large and narrow glumes.

Another important hybrid introduced in the 1930's is *marquis* which has a gradually tapering spike from the base to the tip. It is essentially beardless although some of the lemmas have short, fine awns. Marquis wheat resulted from the cross of *hard red Calcutta* (from India) and *red fife* (from Russia) made in 1892 in Ottawa, Canada (CLARK, 1929). It was introduced into the United States in 1912 and in 1938 was still the most widely used spring wheat cultivar in North America, even though the severe rust epidemic of 1935 eliminated much of its hectarage. In the Vilcanota Valley marquis became a very widespread cultivar in Canchis and Quispicanchis starting in 1932. In the late 1930's, however, marquis was heavily attacked by Puccinia and lost much of the favor it had had several years earlier. Its areal extent retracted and today marquis is found only occasionally below 3,600 m. Above this elevation, however, it is still extensively planted in the Indian communities of southern Canchis. It begins to be dominant 3 kilometers south of the town of Maranganí, and continues to be so virtually to the uppermost level of cultivation at Occobamba, about 3,880-3,900 m above sea level. Around Maranganí it is mixed with candeal and barba negra.

This Canadian wheat is known by the Indians as *mayllko* which means in Quechua 'for a cold place'. However, neither this nor any other wheat is well adapted to the rigorous environment, as testified by a crop failure approximately 1 year in 6, and by the predominance of plasmolyzed grains causing a large loss of nutritive substances. A majority of the caryopses are diminutive and wrinkled. The time required for this wheat to mature is 11 months in this environment. Marquis

[24] In GADEA's classification of the Spanish wheat flora, *espiga negra* (*T. durum coerulescens* Bayle) most closely resembles the barba negra of the Vilcanota Valley. Espiga negra is still grown in Jaén, a province in Andalusia, and close to the port of Cádiz, whence it may have been taken to Peru (GADEA, 1954:344).

has other disadvantages. At first resistant to wheat rust, it is now just as susceptible as any other cultivar, perhaps because other strains of the disease are now abundant to which marquis has little resistance. The spike does not readily shatter when ripe, so it is difficult to thresh by the traditional methods. Considerable unthreshed wheat still remains after the animals have trampled it on a threshing floor, and this residue must be threshed with long-poled flails. As for the grain itself, grinding it with the technology available is so difficult that the peasants sell most of it to bakers in Juliaca and Sicuani where it is esteemed as bread wheat. Wheat is a cash crop for these peasants, and the marquis cultivar has the one great virtue that, with its non-shattering spike and stiff straw, it resists the frequent hail and sleet storms that can completely destroy other cultivars of wheat. Marquis is low yielding here, not disease resistant and requires extra work to thresh and mill. To the Indians, however, the assurance of some harvest for their own use and for sale, is more important than a higher yielding, easier worked crop which also runs a greater risk of failing completely in the event of adverse weather.

Since 1950 other hybrids became increasingly important in other parts of the valley because of their higher yields, shorter growing seasons and better breadmaking qualities. The most common of these cultivars is *florencia-aurora* which is now the preferred wheat between 3,000 m and 3,200 m in the valley. Florencia-aurora is a famous winter wheat developed in Tunesia (PETERSON, 1965:224) where it is a known as *florence aurore* (formerly *cailleux*). The uppermost three lemmas are awned. The grain is considered 'soft' for a *sativum* cultivar and it is quite susceptible to rust, but its yields are high. Other hybrid cultivars are *helvia*, a high yielder grown under irrigation; *mariachi* and *Maria Escobar*, the latter a selection made in 1942 of a hybrid between two Argentine wheats. In the Vilcanota Valley it has passed its popularity because of disease susceptibility. Recently another Peruvian-made hybrid, *H-28*, has become important in Calca and Quispicanchis.

Many peasant farmers in the valley still refuse to adopt the modern hybrid varieties. They note that the caryopsis of most modern hybrids is smaller than the old chumpi wheats and do not realize that actually the amount of gluten is proportionately higher in the modern wheat, because gluten is contained mainly in the exterior coverings of the wheat caryopsis and not in the central endosperm zone. They also think that the smaller spike indicates low yields; but wheat yield does not necessarily depend on the dimensions of the spike. Modern hybrids also are usually beardless and thus more susceptible to bird damage. Another major *Puccinia* epidemic would, however, be the *coup de grace* for many older cultivars in the valley today.

Wheat Cultivation

Peasant farmers use may European methods in wheat cultivation that were common in the sixteenth century in Spain. Plowing is done with oxen and sowing is done broadcast. The seed is generally not sown close, because soils are poor and weeds are not the enemies that they are in North American agriculture. Wheat is irrigated in Canchis to some extent, but elsewhere it is grown on the unirrigated slopes above the flat valley. Not only does irrigated wheat make poor bread, but in most places irrigated land is too valuable to be used for wheat. Harvesting is done with a sickle and cut grain is then gathered into bundles and taken to the threshing floor. Sometimes small bundles are left in the field 2 or 3 days to dry well before being made into larger bundles and removed. Once at the area of the threshing floor, the wheat, if not completely dry, will be exposed to the sun for a time. After the animals have threshed the grain from the husks, the wheat is winnowed. The chaff and stalks are swept up to be used primarily as forage for livestock.

Commerce

Wheat and some flour is an important item of trade within the valley and to towns outside the valley. Sicuani, in the heart of the Canchis wheat-growing area, is by far the largest trade center and much of this wheat goes out to the valley to the northern part of the Altiplano and to the tropical valleys of the east. The lower Vilcanota Valley, however, gets most of its imported wheat from the area around Cuzco and the Pampa de Anta, a very flat surface which lies west of the valley at a considerably higher elevation.

Food

Most mestizos and cholos in the valley eat bread daily, and an estimated 75 percent of the Indians eat it at least once a week. Each year, bread becomes *sine qua non* in the diet of more peasants. Wheat bread is a fairly recent staple in the peasant diet of the valley; elderly informants at Ollantaytambo indicated that in their youth bread was eaten only on special occasions, whereas now it is common. In the nineteenth century (Memoria/Urubamba, 1874:385) and as late as 1927 (DELL, 1927:206), small flat cakes (*biscochos*) as hard as rocks were imported from Urubamba into the valley below 1,500 m. The cost of milling, while low, may have helped make flour for bread uncommon. Today there are bakeries at Urcos, Sicuani, Urubamba, Calca and Quillabamba, as well as outdoor ovens in other towns used especially on All Soul's Day. Because of the low gluten wheat used, the bread is often flat rather than spongy, hard rather than soft. Barley flour is often mixed with wheat and wheat bran is often retained to make whole wheat bread (*pan integral*). Baking bread cannot always be done in the valley where facilities are lacking or fuel is scarce, and other types of wheat foods are prevalent, especially among Indians. *L̂lunka* or *trigo pelado* is wheat which is washed, soaked for about 2 hours, then gently rubbed until the husk is removed and left to air dry. When the grain is older, it is cooked in an alkali solution of wood ashes to facilitate removing the bran, just as maize is processed to make mote. This peeled wheat is eaten with meat, rice and sauces either in soups or stews. Wheat is also sometimes prepared toasted or parched (*kancha*) of puffed (*trigo phata*).

Stipa ichu (R & P) Kunth, *ichu*, is a perennial bunch grass growing in profusion from 2,700 m to 4,300 m in the valley. Ichu grass is remarkably resistant to both drought and frost, and is the most useful wild grass growing in the valley. It is the traditional thatch material on most of the peasant huts in the upper valley and the essential binding material in adobe bricks. Those people who live above 3,700 m make it into rope, and also depend on it as a basic forage for their llamas and sheep. Formerly baskets were made of twisted ichu grass.

Stipa pungens (R & P) Kunth, *iru ichu*, is a flat-leaved grass, lead-coloured, and with small white flowers. It prefers sandy stream banks above 3,700 m. Livestock eat some of it in the rainy season and it is used as thatch and to fashion homemade brooms. An infusion of it is made to treat colds, coughs and pneumonia.

Oryza sativa L., *arroz* or rice, was first planted in Peru in the late sixteenth or early seventeenth century (COBO: I:408). Exactly when rice was brought to the lower Vilcanota Valley is not known for travellers do not mention it. Rice was cultivated for a time in this century by hacendados in La Convención, mainly for local use (BOWMAN, 1916:76). When the railway was built, rice cultivation declined and now is only sporadically grown anywhere in the lower valley. Although rice is a food staple of many peasants in the lower valley, they curiously never took to growing it.

Chusquea scandens Kunth, *kurkur*, is a bamboo which grows on rainy slopes in the valley between 2,300 m and 1,500 m, as a pioneer plant. It is an indicator of sites on which the natural vegetation has been disturbed. The flexible branches of this bamboo are occasionally woven into baskets and mats.
Phragmites communis L., *sok'os* in Quechua and *carrizo* in Spanish, is a grass with thick, hollow stems growing as high as 3 meters. It grows naturally on floodplains in the valley below 3,300 m and is also planted in dooryard gardens because the canes are used as roofing and flooring materials.

Gynerium sagittatum (Aubl.) Beauv., *pintok*, is a streamside grass found below 1,600 m which is used as a construction material in the same way as the above mentioned species. The Machiguenga constructed temporary shelters from this grass when they came down to the riverbank to fish during the dry season; some of these can still be seen near the junction of the Vilcanota and Yanatile Rivers.

Cortaderia quila (nees) Stapf, *niwa* or pampas grass, is a very tall native species found particularly in the Provinces of Urubamba and Calca. It is used here for thatch and is also woven into ropes used to lash beams together. The conspicious panicles are hung beside a doorway to announce chicha for sale.

PALMACEAE

Bactris ciliata, *chonta*, is a small palm tree with exceedingly hard and durable wood. Its ecological position in the valley is in moist sites below 1,200 m. In the Inca period this straight wood was extremely important for making spears and other weapons, maces, war clubs and fish hooks. Chonta wood was probably one of the important trade items from the tropical forest peoples to the Highlands. The intensive exploitation of the wood before and after the Conquest may explain why QUEVEDO (1900:498) made a special point of noting the absence of palms in the lower valley in the early eighteenth century. Chonta wood is still valued in the Highlands for the shafts of tools and as a construction material, but now it must be obtained farther down the valley outside the study area.

ARACEAE

Colocasia esculenta (Linn.) Schott, taro or dasheen, has its cultivation centered in the lower ceja up to an elevation of approximately 2,200 m. Taro tolerates more moisture than other root crops and this partly explains its restricted distribution in the valley to wet climate and/or wet sites. Apparently of Asian origin, little is known about when or how taro came to the New World, much less to the Vilcanota Valley. The Polynesians could have brought it to the west coast of South America on one of their sailing trips; another possibility is that Negroes brought it from Africa to Brazil in the sixteenth century and that it slowly diffused up the Amazon drainage basin. Many forest Indian groups in the upper Amazon tributaries in Peru grow this plant. (TESSMANN, 1930). The Quechua Indians of the lower Vilcanota Valley call taro *papa japonesa* (Japanese potato); such a name suggests that the plant may have been brought from the Far East in historic times. Taro has peltate leaves, clear plant juice and does not flower in the valley. The roundish edible roots are usually boiled and served in thick stew.

Xanthosoma sagittifolium Schott, *uncucha*, is another araceous root crop but differs from taro in its hastate leaves, milky plant juice and New World origin. The upper elevational limit of cultivation is about 2,100 m and it is more common below 1,500 m than is taro, perhaps because it tolerates dry sites better. Apparently it does not flower and reproduction is entirely by cuttings. The important cultivars of uncucha include *lima* with white flesh and a smallish root; *kello*, with yellow flesh; *platillo*, with a flattish root; and *waskamanuku*, with a very large root but containing many calcium oxalide crystals.

Cuttings of uncucha are planted from October to December and the crop is harvested in July and August. It is often planted on recently cleared lands while tree crops are maturing. As a crop, uncucha has an inestimable advantage over manioc in that it can be stored for long periods outside the ground without becoming unpalatable. However, most peasants prefer to eat manioc rather than uncucha, because calcium oxalide crystals in the latter leave an unpleasant sensation in the mouth. The starch of uncucha is also considered to be less digestible than that of manioc. Uncucha is usually prepared boiled in a thick stew. It is less commonly roasted. The roots are occasionally dried, but not, as O.F. COOK (1916c:341) indicated, by making 'chuño'; they are simply dried without the freezing implied by the word 'chuño'.

BROMELIACEAE

Tillandsia usneoides L., *wik'ontoy* or Spanish moss, is a moss-like epiphyte that is collected from rocks and bushes between 2,400 m and 3,400 m and used to stuff mattresses and pillows and for packing material around earthenware and other fragile objects.

Puya spp., *achupalla*, is a stiff-leaved wild plant 1 meter high with a woody base. Perhaps the most common of several closely related species of this genus found in the upper valley is *P. longistyla*. Much *Puya* grows on rocky south-facing slopes at elevations between 2,800 m and 4,100 m. Its most intensive human use is at the highest elevations, above 3,700 m, where the woody inflorescences are used as beam supports for the low grass huts constructed in that area (Fig. 29), and where the rhizome-like underground part is used as fuel.

Ananas comosus Marr., *piña* or pineapple may not have been found in the valley in pre-Conquest times. GARCILASO DE LA VEGA does not mention it, although VARGAS (1962:111) identified as a pineapple a suggestive – but hardly unequivocal – painting on an Inca wooden vase from Cuzco. Today it is cultivated mainly below 1,300 m, but occasionally will be seen as high as 1,650 m. Pineapple is a common market item in the lower valley that also reaches the larger towns in the Sierra. Most production is on a casual basis, in dooryard gardens around the peasant huts.

JUNCACEAE

Juncus spp., *chiwa*, is a rush that forms mats in certain poorly drained areas of the temperate valley. Cattle graze on it, especially in the dry season when it is one of the few green plants available.

LILIACEAE

Allium sativum L., *ajo* or garlic, was one of the Old World crop plants more enthusiastically accepted by the indigenous population. ACOSTA (1962:236) claimed that the natives valued garlic 'above all the roots of Europe and hold it as a fruit of great force.' COBO (I:415) related that the the Sierra Indians made greater use of garlic than of any other European produce. Although garlic is still highly regarded as a condiment, onions are now as popular if not more so in the peasant diet. Some garlic is grown for local consumption in the valley, however, most of it consumed in the Department of Cuzco comes from the Arequipa area.

Allium cepa L., *cebolla* or onion, was introduced into Peru in the sixteenth century but no one knows when it reached this part of the Sierra. COBO (I:415) wrote that onions were restricted to the coastal plains because of their low resistance to cold; if this were true, then the onion could not have been common in the valley until later. Whether or not it was the separate introduction of a new, hardier kind of onion that made possible its spread into the Highlands is open to question. In any case, today in the valley, the onion is cultivated as a winter crop up to 3,650 m and is well adjusted to the low temperatures of this season. Only one basic kind is grown, *común*, the bulbs of which are noticeably smaller than those raised in the mid-latitudes. Onion flowers are used ornamentally on All Souls Day. An intensive onion-growing center is the Indian community of Uscupata, 4 kilometers from Sicuani, where onions are harvested to be sold in the Sicuani Sunday market.

Lilium candidum L., *azucena blanca* or Madonna lily, is an Old World ornamental introduced in colonial times to Peru (COBO:I:412). The blooms are often grown to be sold and this lily is one of the most popular flowers in the valley.

Nothoscordum andicola Kunth, *chullkus*, is a wild plant yielding edible bulbs tasting somewhat like insipid garlic. These bulbs have been a meager source of nourishment in Southern Peru since pre-Columbian times (COBO; I:172). They are still collected from the hills in Canchis in August and September, after frost has dried the leaves. Sometimes the bulbs are dug up as the land is plowed. Eaten boiled, the bulbs are valued especially when other food is scarce and they are the most appreciated source of non-cultivated food in the zone above 3,400 m.

Smilax officinalis L., *sarsaparilla*, has been one of the most highly valued medicinal plants in the valley. Although it is a native species, it has only a Spanish name today and its use as a remedy for liver and kidney ailments may postdate the Conquest (ACOSTA, 1962:191). This plant grows in areas below 2,500 m; it is collected especially from the forested hills of the upper ceja and sold in the Cuzco market.

DIOSCOREACEAE

Dioscorea discolor, *maguma*, is a new world yam that grows on a very limited scale in the valley below 1,000 m. The Quechua peasants adopted this cultigen from the Machiguenga forest Indians who still cultivate it in their chacras in the Urubamba Valley. In addition to its native name, it is also called *papa silvestre*, because it is the tropical rootcrop known in the valley that most closely resembles the true

Fig. 49. Extraction of fibers from the leaf of the *paka* (*Furcraea andina*) plant to be made into a tough rope. Near Santa Teresa, 1,530 m.

potato.

AMARYLLIDACEAE

Two different species, both called *pakpa* in Quechua, are found in the valley: *Agave americana* and *Furcraea andina*. The fibers of both are known indiscriminately as *cabuya* and have been used for essentially the same purposes. *Furcraea andina* Trel. is found between 2,700 and 2,500 m., then reappears at 1,600 m and extends down the valley to the Yanatile River. The gap between 2,500 m and 1,600 m is due to excessive moisture which hampers the growth of the plant. *Furcraea* appears to be in its native habitat in the valley, but it is occasionally also planted in hedgerows. During the Inca period, the fiber was used extensively to fashion suspension cables for bridges. Quarried stones were conveyed from the quarry by means of great cables of *cabuya* and leather (MARKHAM, 1968:161), and even clothing was made from the fiber of this plant in communities that did not have llama fleece (SANTILLÁN 1968:124). The fibers were a tribute item in the colonial period, and rope is still made from the fibers today.

To make rope fibers, each leaf (*penca*) is broken into 8 or 10 strips (Fig. 49). Two sticks of wood about 22 centimeters long are tied together and held vertically braced against a rock with the left foot. The left hand presses the sticks together while the right hand pulls the leaf strip through the two sticks, scraping off the pulp. Since the worker can put only half the length of the strip through the sticks, each strip is then reversed. Each half-piece is put through the stick five times to completely clean it of pulpy material. The pulpy residue is then discarded and the ends of the fibers are tied together and hung up to dry before being woven into rope.

Agave americana L., *maguey*, is not a wild plant in the valley as is the previously discussed species. Maguey may have been brought from Mexico before the end of the sixteenth century, although its similarities with *Furcraea* make certainty difficult. The fibers were extracted from agave by much the same process as from the native plant, and were made into rope to be used as bridge cables. In the colonial period, sandals (*ojotas*) made of cabuya formed an item of tribute from many Indian communities (JIMÉNEZ DE LA ESPADA :I:195). The sandals continued to be made until the 1930's when old truck tires began to be cut up and fashioned into footwater for the peasant class.

The bluish-green agave plant is found in the valley above 2,700 m and is particularly common in the landscape of Quispicanchis and northern Canchis. Rows of agave outline most of the fields near the towns of Combapata, Andahuaylillas and Tinta. This use as a boundary indicator is probably the most important one that the plant now serves. Rope is no longer commonly made in the upper valley. Reportedly the leaves are still occasionally used to make lye and to make soap for washing clothes. The dried leaves are burned as fuel and, as a bonus, the ashes yield a fertilizer rich in potash. In this tree-poor region, the fruiting stalks of agave are in demand for beams and poles in rural constructions. In some parts of the Andes (RODRIGUEZ SANDOVAL, 1949:75), the juice (*miel*) from the stem is made into a fermented beverage like pulque, but this is apparently unknown in the Vilcanota Valley.

MUSACEAE

Musa spp, *plátano* or banana, a very efficient food producer, is cultivated in virtually all areas of the valley where it will tolerate the climate. Three of the presently recognized species of edible bananas grow in the valley: *M. sapientum*, the common banana; *M. cavendishii*, the dwarf Cavendish banana; and *M. paradisiaca*, the plantain. Among the sixteenth century chroniclers, ACOSTA (1962:179) and GARCILASO DE LA VEGA (1960:152) both asserted that the banana was indigenous to the New World. More recently, there are scholars such as JEFFREYS (1963) who, while acknowledging its Old World origin, tend to support a pre-Columbian introduction to the New World. No strong lines of evidence support a pre-Columbian introduction into the Vilcanota Valley. There are no seed bananas, few native names for the fruit, no prehistoric archeological evidence, and it has no strong place in the domestic economy of the inhabitants. The earliest written suggestion we have of the plant in the Cuzco area is *c*. 1551 when CIEZA

DE LEON (1883:339) noted it in the Cuzco eastern valleys. It is not clear what kind of banana was present in the valley in the sixteenth century; we can be sure, however, that it was neither *gros michel* nor *cavendish*, the two popular cultivars today.

M. sapientum L.: The outstanding cultivar of the common banana in the valley is *seda*, known in some other countries as *gros michel*. It was taken to Jamaica in 1836 from Martinique where it appeared in the nineteenth century (SIMMONDS, 1959:78); thus it has probably grown in the valley less than 100 years. Seda attains a height of 4-6 meters and sends out suckers which form large clumps. The bunches are curved and the tapering fruit, 15-22 centimeters long, has a bright yellow, thick skin. Seda banana is very susceptible to Panama disease, is not tolerant of stoney soils, and cannot withstand much cold. This cultivar has its upper limit at 1,840 m on the river bottom; it does not, however, become important until near 1,500 m, and does not grow on the lateral slopes above 1,350 m elevation. Seda is the only significant commercial variety grown and large quantities are shipped up the valley to the Highlands. Among the local folk, it is eaten fresh or fried. Other cultivars of this species in the valley are *manzana, chino, pera, imperial, plátano de isla* – all have individual flavors – but are by no means grown on all chacras.

M. cavendishii Lambert, *plátano enano* or dwarf cavendish banana, originated in South China whence it was taken to Mauritius in 1826; the South American material came from the Canary Islands in the nineteenth century (SIMMONDS, 1959:80). Thus, although the date of introduction into the valley cannot be pinpointed, it may be presumed to be after 1850. Enano plants grow $2\frac{1}{2}$-3 meters tall with broad but short leaves. The fruit is 12-15 centimeters long and when fully ripe, it has a light yellowish-green skin and orange flesh. The fruit, unlike the plant itself, is delicate and sensitive to bruising and low temperatures. Enano banana does well on very poor soils, and one often sees it growing out of rock rubble on 50 percent slopes. Unlike seda, it is resistant to Panama disease. Also, enano is hardier than any other cultivar grown: the extreme upper limit in the valley is 2,200 m and the effective upper limit is 2,050 m. At elevation 1,820 m enano banana is seen growing profusely on cliff-hanging chacras near the river; this pattern continues down until about 1,400 m where other cultivars replace it. Few cultivated plants have such a well-defined elevational niche. That part of the harvest destined for sale is wrapped in straw and brought to the transportation head on the backs of burros.

M. paradisiaca L., the plantain, is of minor importance in the lower valley, because the inhabitants, of highland origin, have not integrated this tropical staple into their diet. However, several kinds are known: *kulli* (Jamaica red) used for frying; *hartabellaco*, a very large fruit which is roasted; and the *waik'u* which is boiled.

Banana rhizomes of whatever species or cultivar are planted about 2 meters apart and with very little care are ready for harvest from 10 to 16 months later.

The same plant may be kept up to 20 years or more, although 10 years is a more typical period. The place of the banana in the diet of peasants in the lower valley is less than one might expect from the ubiquity of the fruit. It is largely a cash crop for the peasants who find a market in Cuzco and other Highland towns. Furthermore, the great majority of farmers are serranos who have Highland dietary preferences and who greatly prefer the potato over the banana. Bananas are generally eaten raw and fresh. Although they are sometimes fried, boiled or roasted, this is less common in the Vilcanota Valley than in other lowland tropical environments. Occasionally bananas are sun-dried. Other parts of the plant have some uses: the leaves are used as thatch and as a cooking fuel, and the outer 'bark' from the false trunk is made into containers for transporting coca.

ZINGIBERACEAE

Curcuma longa L., turmeric, also known as *palillo chuncho* in the valley, is cultivated for its root which yields a saffron-yellow dye used to color food. In Peru it is generally thought to be of Amazonian origin, but it is actually a South Asian plant which was brought to Peru after the Conquest and was subsequently cultivated by forest Indians. The forest Indians, in turn, probably traded it to the Highland Indians living in the lower Vilcanota Valley, although proof of this transfer is lacking. Turmeric is cultivated commercially by some peasants and in 1956, 138 hectares in the lower valley were in this crop (according to unpublished statistics from the Ministry of Agriculture, Cuzco branch). It yields an average of 700 kg/ha and is sold dried to wholesalers who distribute it throughout the country as a coloring agent.

CANNACEAE

Canna edulis Ker., *achira*, is a subtropical domesticated crop with a starchy rhizome. It has a very long history of use on the Peruvian coast and archeological evidence is fairly abundant (TOWLE, 1961:34). Achira may have been important in the Vilcanota Valley in the past, but today it is cultivated only sparingly in the area in which it can grow well, that is, below 2,100 m. It is occasionally seen on hillside chacras where it forms field borders. In the lower valley the rhizomes are boiled. The most intensive center of cultivation of this plant in the Peruvian Sierra lies on the banks of the Apurímac River in the Province of Paruro (GADE, 1966). Its use here is closely tied to the Corpus Christi festival in the month of May (or June) which is celebrated in Cuzco as well as in certain towns in the valley such as Urcos, Sicuani and Combapata. The baked rhizome of achira seen in these towns during this festival all come from tiny villages on the Apurímac River and are not grown in the Vilcanota Valley itself.

ORCHIDACEAE

Vanilla sp. is a cultivated vine with a valuable pod which is dried and used as flavoring. Vanilla apparently no longer grows within the study area, though it was cultivated in the lower valley in the nineteenth century (CARRASCO, 1847:232). There is some indirect suggestion that it was brought up the Urubamba River drainage by forest Indians who may have traded it to the Highland Indians.

Cultural-Geographical Analysis of the Individual Floristic Components: Dicots

PIPERACEAE

Below 2,400 m several shrubby wild plants of the genus *Piper* are collected for their medicinal properties. The most important species are *Piper angustifolium* R & P, *P. elongatum* Vahl. and *P. sinuclasum* Trel., but all the species are given the common name of *moko moko*. The leaves have soothing properties and they are dried, powdered and applied to the wounds of man and beast. Infusions of the leaf are also made for kidney ailments.

SALICACEAE

Salix Humboldtiana Willd., *sauce* or willow, is a small tree that is believed native to Mediterranean Chile. The Incas may have brought it to Peru, and planted slips of this willow along the banks of the Vilcanota, not only to help prevent river overflow but also to supply wood and osiers. Many indigenous bridges were made of this willow, e.g. the one on which MARCOY (1875) crossed the Combapata River in the 1870's; indeed, willows may have been planted specifically to provide the raw materials for these bridges which were so important in the Inca transportation system. All through colonial times and until the twentieth century, willow branches were used to make bridges. The inhabitants of Yucay and Huayllabamba traditionally made wicker baskets (GONZALEZ 1833:51), an activity which is much reduced today. However, *puktus* (round baskets for carrying fruit) are still made in these towns and today the largest concentration of willow trees in the valley are along stream courses or irrigation ditches in the Huayllabamba and Yucay areas.

JUGLANDACEAE

Juglans neotropica L., *nogal* or walnut, is a New World species associated with man, especially in Urubamba and Calca Provinces, but cannot be called truly domesticated. It is a tree native to the ceja de la montaña now rare in the wild, that has been planted at higher elevations. COBO (I:257) noted that the nuts were sparingly eaten, but said nothing about using the leaves for dyeing. Today walnut brown is the only plant pigment still commonly used to dye textiles (mainly ponchos). Branches are collected from trees which grow along roads in the Urquillos Yucay-Urubamba area and are traded to communities above the rim of the valley. The tradition of dyeing with walnut leaves in this area may be related to the textile factories (*obrajes*) established there in the early seventeenth century (GADE, 1968). The branches and leaves are crushed with a stone and then boiled in a large pot for as many as 10 hours. In addition, the leaves, bark and nutshells have minor medicinal use, but the nuts are largely ignored.

BETULACEAE

Alnus jorullensis H.B.K. *ramram* (or Hispanicized to *lambrán*) is a streamside native alder tree of the upper valley, seen most commonly in Urubamba and Calca. The wood of this tree makes excellent charcoal, and it is said that during the War of Independence lambrán charcoal was widely used in gunpowder manufacture. It is now used as a general firewood, for construction beams, yokes, and as a cabinet wood, but large specimens are today quite rare in the valley.

MORACEAE

Castilloa elastica L., *caucho*, a latex-bearing tree, has been exploited as an alternative source of rubber. The true rubber tree, *Hevea brasiliensis*, is not found beyond 300 meters above sea level and does not occur within 100 kilometers of the study area. Neither the quality nor the yield of *Castilloa* latex is as high as *Hevea* latex (*jebe fino*). *Castilloa* sparingly grows in the wild downvalley from about 900 meters below Echarati. Late in the nineteenth century and early in the twentieth, both lowland and Highland Indians were forced to collect *Castilloa* – some on the northern fringe of the study area – and as a result of this tapping, the species may be much rarer than it was.

Cecropia peltata L., *torok*, a tropical weed tree, begins in the valley at an elevation of 2,350 m. It is considered to be an indicator of ecological disturbance and commonly forms a second growth (*purma*). Some people split the trunk and use it in part of their hut construction, not because it is a strong wood, but because it is easily cut and trees are readily accessible.

Artocarpus communis Forst., *árbol del pan* or breadfruit, is a Southeast Asiatic native tree cultivated for its edible fruit; its status as a curiosity in the valley may suggest a twentieth century introduction.

Morus nigra L., one of the two Old World cultivated mulberries, was introduced into the lower valley in the nineteenth century as part of an attempt to start a silk industry (CARRASCO, 1847:232). In the 1850's CASTELNAU (1851:287) observed more than 100,000 young mulberry trees at Hacienda Echarati. Although the trees themselves were successful, exportable silk was, as far as can be determined, never produced. Mulberry has now volunteered below 1,000 m in La Convención; it is collected from the roadsides and used as a firewood.

URTICACEAE

Urtica urens L., *mula kisa*, is a nettle used in native medicine to help cure lumbago by rubbing the leaves on the sore spots.

LORANTHACEAE

Phrygilanthus ellipticus, paukarkay, a parasitic shrub found in crop fields, is regarded as a noxious pest.

POLYGONACEAE

Rumex crispus L., *wacha*, an Old World weed is now common in the Vilcanota Valley; the leaves and stalks are fed to cattle.

Rumex cuneifolius Campt., *llake*, is a weedy plant with yellow roots which are used to 'purify' the blood.

Triplaris peruviana R & P, *palo santo*, is a tropical tree growing wild in the lower valley with very hard wood used in construction.

CHENOPODIACEAE

Chenopodium Quinoa Wild., *quinoa*, a domesticated herbaceous plant, attains a height of 2 meters, and has broad leaves increasing in size and number toward the base of the stalk. The root is simple, generally braided and easily pulled up. The inflorescence is a panicle and the fruit is an achene which encloses only one tiny seed. The cuticle surrounding the seed contains saponin and this must be washed away before the seed is palatable.

The peasants classify quinoa on the basis of seed color and size. Three kinds are grown in the Vilcanota Valley: *yurak*, with large white seeds; *kami*, with yellow seeds and yellowish stalk, leaves and fruit; and *puka*, with red stalks, fruits and yellow seeds. The latter two kinds are said to contain a higher percentage of saponin than does the first which explains why yurak is preferred. Quinoa is apparently a Hispanicized from of the original Quechua word *kiuna* and today

Fig. 50. The fruits of quinoa are removed from the stalk and placed to dry. Near Tinta. Elevation 3,420 m.

even Indians use the corrupted term. The Aymara people call the plant *huppa*, but the Spaniards first noted the crop among the Quechua and so adopted their name for it.

Quinoa is sown broadcast in the months of September, October or November. When the plants reach a height of 6-10 centimeters, they are already extremely resistant to drought, so moisture is critical only during the germination stage. Few pests bother quinoa and farmers give little attention to weeding it. The crop is harvested 150-180 days after sowing. The stalks are often pulled out of the ground by hand rather than cut at the base with a sickle. One apparent reason for the survival of this pre-Conquest harvesting method is that, unlike most grains, quinoa seed shatters readily when ripe, and pulling jars the stalk less than cutting it. By extirpating the whole plant, nothing remains in the field to provide organic matter to the soil. The stalks and panicles are dried in the sun for several days (Fig. 50), before being threshed by flailing sticks; rarely if ever is quinoa threshed by animal hooves in the valley.

The distribution of quinoa is wide: from 2,770 m to 3,800 m. In no place in the Vilcanota Valley is it grown as a basic crop, occupying unmixed fields. Curiously quinoa is least seen in the harsh microthermal environment of southern Canchis where one would most expect to find it cultivated. The most common reason given for its scarcity here is that the people do not know how to prepare it for consumption; this suggests that wheat and barley have replaced it here in recent generations. Farther north in the valley, in the maize growing section, quinoa is

Fig. 51. Chola selling, among other products, *llipta* (small, roundish balls in left centre) made from the ashes of quinoa stalks and used in coca chewing. At Pisac market.

often sown as a living fence around a maize field, a planting pattern which goes back to Inca times (GARCILASO DE LA VEGA, 1960:227). The Indians claim this 'defends' the maize from winds, frosts, and animals. Quinoa may also be sown in clusters within a maize field or individually dispersed among corn plants.

Only where wheat and barley do not thrive, as on the Altiplano (south of La Raya Pass), is quinoa still the basis of subsistence. The loss of quinoa as a significant crop in the Vilcanota Valley is lamentable in view of the advantages of this plant: few pests, high food value, a variety of uses for most of the plant, and ability to produce despite low rainfall and temperatures and poor soil conditions.

Not only does quinoa have a higher protein content than grain (WHITE et al., 1955:534), but the quality of its protein is very high. After they have been thoroughly washed, the seeds are made into a mush, prepared by first toasting and then boiling them. Often sugar is added. As late as the mid-nineteenth century (GONZÁLEZ, 1833:61), quinoa mush (*api*) was important in the diet of Cuzco town dwellers and rural folk alike. Another way to prepare quinoa was to grind the seeds and to make them into a dough which was allowed to harden into a cookie. Known as *kispiño*, this was a common food taken on journeys because it would last several months without spoiling. Quinoa chicha was formerly made, but now it is prepared in the valley only for certain ceremonial occasions, for example, the Feast of Corpus Christi. Before the plant matures, the tender leaves are eaten as a salad (*llipecha*), furnishing vitamins. Quinoa stalks are macerated, dried, and burned; the ashes are then collected and formed into a ball, the size of a golf ball, a bit of which is chewed with coca leaves to extract the alkaloid in the leaf (Fig. 51). The stalk is also valued as a fuel which produces a quick, hot fire in an area where fuel of any kind is scarce.

Chenopodium pallidicaule Aellen, *kañiwa*, is much smaller (30-50 centimeters) and more branching than quinoa. Kañiwa is grown sparingly above 3,800 m in the valley beginning at the lower edge of the puna and extending to La Raya Pass (4,340 m). It commonly grows in small patches on a garbage heap or in an abandoned llama pen, land having been fertilized by the droppings. Of all cultivated plants in the valley, kañiwa requires the least amount of heat for successful growth. Planted in the months of November, December and January, it matures within 120 days. It is threshed like quinoa, with a stick, and the seeds are used in much the same way as quinoa (GADE, 1970d). Kañiwa has an even higher protein content (WHITE, 1955:532) than does quinoa, but less calcium. This may explain the preference for quinoa ashes as the accompanying lime for coca chewing.

Chenopodium ambrosioides L., *paik'o*, an herbaceous plant with a very strong odor grows as a ruderal at elevations above about 2,600 m. It has long been part of native medicine (COBO: I:179) with the seeds used against worms and the leaves to cure stomach disorders.

Chenopodium petiolare L., *llipecha*, a weed in grain fields, is valued for its edible leaves that are consumed boiled.

AMARANTHACEAE

Amaranthus caudatus L., *kiwicha*, is a domesticated herbareous species yielding many tiny black seeds from a large panicle. The seeds are toasted and the leaves are eaten as potherbs. This plant is cultivated very sparingly from 2,500 m to 3,000 m and perhaps it never was important in the valley. Field observations from the Ayacucho region of Peru (several hundred kilometers north of the valley) indicate that kiwicha has maintained itself in fairly warm, poorly watered areas where it thrives on unirrigated sites. In the valley warm areas are also wet and thus kiwicha has no clear ecological position here.

Amaranthus quitensis H.B.K., *hatak'o*, includes a weedy group of amaranths that is probably the most common adventive in the Vilcanota Valley. This weed, growing in all kinds of fields and at all elevations, is valued for its leaves which are boiled and eaten as a potherb. Also within this species group is at least one cultivated plant known as *kiwillo*, which yields edible seeds, lime from the ash (for coca chewing) and from some plants a red dye (*caramina*).

BASELLACEAE

Ullucus tuberosus Lozano, *ullucu* or *lisas*, is a domesticated tuberous plant with small magenta-colored flowers and a reddish tinge to both the simple leaves and to the stalk. Ullucu does not set seed in the valley. The edible tubers come in a wider variety of shapes and colors than does either añu or oca. Small, roundish tubers may be light yellow, golden yellow, lilac, light green or light or dark magenta. Long cylindrical tubers may have any of the above colors, but are most notably either yellowish or green with magenta speckles. Ullucu is grown from 3,160 m to 3,980 m in fields with other tubers. Below 3,150 m it is cultivated sparingly on high hillsides. Ullucu is considered the most resistant of all Highland tubers, perhaps because the plant itself hugs the ground more closely than do the others. Unlike oca and añu, ullucu tubers are an acceptable food among all of the social classes of the area. This may be partly because ullucu requires no special preparation. It is always used fresh, as thin, narrow slices cooked in thick soups. Dehydrating the ullucu tuber is known in the valley but apparently no longer done. Ullucu cannot be satisfactorily baked as can the potato, nor is it laid out to sweeten in the sun as are oca and añu.

NYCTAGINACEAE

Bougainvillaea spectabilis Willd., *papelillo*, native to Brazil, was probably introduced to the valley in post-Conquest times. The attractive blossoms make it one of the most often cultivated ornamental shrubs in the valley below 1,800 m.

Mirabilis prostrata (R & P) Heimerl, *wilko*, a twining weed growing in grain fields is fed to livestock. The seeds are said to be poisonous to human beings.

BERBERIDACEAE

Berberis boliviana Lechl., *chekche* or barberry, is a small thorny shrub very widespread in the valley above 2,800 m. In some places it is planted on top of adobe walls as a living fence. Although it plays host in the wheat rust cycle, the peasants do not seem to recognize its role in spreading this disease.

ANNONACEAE

Annona cherimolia Mill., *cherimoya*, a native Andean fruit tree yields an apple-sized fruit with a delicious flavor. Each of the united carpels of the fruit contains a large black seed which is surrounded by a sweet, juicy, white flesh. The pre-Conquest presence of cherimoya is well established in Peruvian culture history (TOWLE, 1961:38-9). The cherimoya tree is cultivated below 1,800 m in the valley and, in spite of its highly valued fruit, has not been commercialized in the zone to any notable extent. It has the status as a semi-luxury food that appears in the larger markets from March to July. One sees it mainly as a dooryard tree or as a field border tree in coca plantings.

Annona muricata L., *masasamba* or soursop, while planted in the same localities as cherimoya, is inferior to its relative in flavor, less common in dooryards and almost never seen in peasants markets.

LAURACEAE

Aniba Puchury minor (Mart. [Metz]) *pucherí*, is a lauraceous of the lower valley with an aromatic bark and a seed used as a purge.

Persea americana Mill., *palta* or avocado, is clearly New World in origin, and there is abundant archeological evidence of the plant on the Peruvian Coast from pre-Conquest times (TOWLE, 1961:41). The avocado is grown as a dooryard tree down valley from 1,700 m. The tree is appreciated for its oily fruit, which is sold in peasant markets although it is much smaller than the improved varieties grown on the Coast.

PAPAVERACEAE

Argemone mexicana L., *k'arwinchu* Mexican poppy, is a weed growing in grain fields throughout the valley. It may have been introduced from Mexico, perhaps on purpose but more likely accidentally, in early colonial times. Formerly the seeds are said to have been used as a source of yellow dye. Today the seed is used as a purge and in stronger doses it is said to have a hallucinogenic effect.

Bocconia frutescens L., *yanalí*, is a wild shrub of the upper valley with a root which is crushed and used as a poultice to heal wounds. Informants claim that the root was formerly used also as a source of yellow dye.

CRUCIFERAE

Capsella bursa pastoris L., *bolsa bolsa* or shepherd's purse, is an Old World herbaceous weed found throughout the upper valley and eaten by livestock.

Nasturtium officinale R.Br., *ok'oruro* or water cress, is an Old World aquatic plant common in irrigation ditches throughout the upper valley. COBO (I:415) mentioned that the leaves and stems were eaten raw in the early seventeenth century, as they are still eaten today as a salad. This plant contains iodine and wittingly or not helps combat goiter, a prevalent disease in the Peruvian Sierra.

Brassica oleracea L., cabbage, is the only domesticated brassica of any importance in the peasant horticulture in the valley. Two kinds have been grown in the valley since colonial times when they were introduced from Spain: head cabbage (*B. oleracea* var. *capitato*) and kale (*B. oleracea* var. *acephala*) (COBO, I:416). Kale is rare today, but this kind of cabbage may have been more important than head cabbage (*repollo*) before the twentieth century, as it was in Europe. The kind that the peasants grow mostly today is repollo. It is planted in fields after the *maway* (early) potato harvest in January or after the maize harvest in May. If the cabbage is started in May, it must be irrigated, but it is appreciated as one of the few crops that can survive the freezing June and July temperatures in the upper valley. Some cholo farmers are now starting to grow cabbage on a highly commercial basis in the Urubamba area, for they are able to sell it readily in the city of Cuzco and in the lower valley.

Brassica napus D.C., *nabo* or turnip, was introduced by the Spaniards into the Sierra in the sixteenth century. COBO (I:415) stated that the Indians did not like to grow turnips because they had their own native roots and thus the turnips that did appear in markets had been volunteers in grain fields. Today the fleshy root turnip is cultivated on a miniscule basis on a few hillside chacras below 2,500 m. Rape, the oil-yielding variety of turnip, may have been introduced but it is not cultivated today.

Brassica campestris L., wild mustard, is a very common and the most useful weed in the valley. Locally called *yuyo* in Quechua and *nabo* (even though it is not the true turnip as discussed above) in Spanish, it is the most conspicuous Old World non-domesticate in the valley between 2,400 m and 3,900 m. The plant is usually less than 1 meter high and has fleshy dentate leaves and bright yellow flowers. It heavily invades crop fields, particularly those of wheat and barley, and because Indians do not sow thickly, a field may contain more yuyo plants than wheat plants. Grain and yuyo seeds become mixed and are harvested and then later sown together; in fact, this plant may have been accidentally introduced into the valley directly or indirectly from Spain mixed with crop seed.

Unlike some brassica varieties, it has no fleshy root that might be edible nor oily seeds that might be pressed. Yuyo does have, however, edible leaves used as a potherb and herein lies its great value (GADE, 1972d). During October, November and December food is scarce, the supplies of the previous year having been nearly consumed, and the new crop yet far from ready for harvest. It is at this time that the leaves are eaten, usually prepared as a dish called *yuyo wacha*. Only the tender young leaves are eaten by the Indians. By the time the plant has flowered, it is fed to livestock. Undoubtedly this potherb provides significant vitamins to country people who otherwise might not get them.

Raphanus sativus L., *rabano* or radish, is a common weed which invades maize and other crop fields below 3,000 m. It is not planted as a crop in the valley, although the small roots are sometimes eaten. The leaves are also eaten when young, but are not as well liked as yuyo leaves. They are also fed to livestock.

ESCALLONIACEAE

Escallonia resinosa R & P, *chachacomo*, is a native wild tree which grows to heights of 8-15 meters. It has simple leaves, flowers in racimes, reddish wood and a twisted trunk. In the Vilcanota Valley, chachacomo is occasionally found growing on steep slopes between 2,700 m and 3,400 m. Few large specimens remain probably because of the long history of cutting. The wood, because of its durability, is used to make household objects (such as spoons), mattock handles and agricultural plows. It is also used as firewood and for making charcoal. The leaves are eaten by a lepidopteran larva (*Eosinga metardaris*) which the Indians relish as a boiled tidbit.

CUNONIACEAE

Weinmania bifida Poepp., *wichullu*, a wild shrub of the ceja zone, has great prestige as a diuretic.

ROSACEAE

Rosa centifolia L., *rosa*, is an introduced Old World ornamental plant found associated with many peasant dwellings in the valley above 2,500 m.

Fragaria chiloensis L., the Chilean strawberry, was brought to Cuzco in the year 1557 according to GARCILASO DE LA VEGA (1960:373). Although the chronicler GARCILASO gave the name *chili* to the fruit, in the indigenous Araucanian language it was known as *quellen* and in the southern Sierra today it is called *frutilla*. The Araucanian practice (COBO: I:157) of making chicha from the berries was not adopted in the valley, but mashed strawberries are added to maize chicha to make a drink called *frutillada*. The fruit of this strawberry is pinkish in color rather than deep red and has a somewhat insipid taste, but all social classes consider it to be a delicacy. The root of the strawberry plant is used in folk medicine and it is traded to Indians on the puna who use it as a blood purifier.

Fragaria chiloensis is extensively cultivated in the Urubamba-Yucay-Huayllabamba area, where an estimated 65 hectares are in strawberries. A noteworthy concentration is at Yucay on a small alluvial fan, all in the hands of cholos. Here strawberry runners are planted in furrows; a job, for some reason, done at night. Care is taken to irrigate the berry plants about 10 times a year and to weed them every 2 months during the first 2 years. The plant flowers in July and August and the harvest begins in November and ends in February. During this harvest period, the berries are picked twice a week and are placed in wicker baskets (*puktus*) which hold from 400 to 500 strawberries. Thirty puktus a week are harvested from 1 hectare; well-fertilized and well-weeded patches yield 45 puktus. In the Cuzco market, each basket is sold for about 50 soles; thus 1 hectare can bring as much as $US 2,000 making strawberry growing a highly profitable kind of horticulture. Its importance and commercialization is based on the demands of a large and nearby market, the city of Cuzco.

Fragaria vesca L., the common wild strawberry, is called *fresa* by the country people to distinguish this species from the cultivated Chilean strawberry. The fruit and root of this plant are collected from abandoned terraces and along paths and railroad rights of way.

Rubus roseus Poir., *kari kari*, is a berry-yielding bush that grows wild in disturbed locations below 2,600 m. The people eat the sweet fruits.

Prunus persica Batsch., *durazno* or peach, is an important fruit tree in the upper valley. It was introduced by the Spaniards in the sixteenth century, and the first time COBO (I:399) saw peaches anywhere in Peru, they were in the Department of Cuzco in 1609. The two most common kinds of peaches in the valley today are *blanquillo* and *koriñana* both of which may have been originally brought to this area in colonial times. The former, a clingstone peach, is small, white and insipid in taste; nevertheless the peasants like it very much and grow it near their dwellings to sell in abundant quantities at local markets, particularly in February. Another old peach cultivar is koriñawa, a yellow-colored freestone peach; it is less grown than the former.

Prunus domestica L., *uñuela* or plum, is another stone fruit dating from the colonial period; COBO (I:402) mentioned it as the most abundant fruit in the Yucay area in the early seventeenth century. Today other fruits appear to have surpassed the plum in popularity. A small black plum that yields well in Urubamba and Calca may be the original stock.

Prunus serotina Ehrh. ssp. *capuli, capulí*, is a bird cherry that was probably introduced from Mexico after the Conquest. Its name corresponds to the Nahuatl word *capulquauitl* and ACOSTA (1962:185) noted seeing capulí 'only in Mexico and in no other country'. It has no pre-Conquest associations in Peru and there is no archeological or documentary evidence of its presence in the Central Andes before the sixteenth century.

Today capulí is a very conspicuous component of the Vilcanota landscape between 2,600 m and 3,500 m, and especially common between 2,800 m and 3,000 m. Capulí trees grow on the edges of maize fields and serve as hedgerows and boundary indicators. The fruits begin to ripen in December and birds compete with the peasants for them. Cholas collect the fruit in round wicker baskets and sell them at markets in this unit. At 10 soles per basket, capulí is the cheapest fruit in the upper valley. The fruit has a large seed and the meager flesh has a rather flat but not unpleasant taste. If it were improved by plant breeding, capulí – so it would seem – could be transformed into a fruit the size and flavor of the Old World cherry. Such a change, however, would necessitate greater attention to cultivation than is presently practiced. The trees are not pruned, sprayed, nor are they irrigated except in so far as they derive moisture from being close to the irrigation canals. Because of its resistance to water rot, the wood of this tree has long been used to make mill wheels; also it is used for firewood and, formerly, to make charcoal.

Pyrus communis L., *pera* or pear, is a much appreciated Old World fruit tree in the upper valley. The ancestral pear introduction, dating from colonial times, is probably the *perilla*, a very small pear that begins to yield in large quantity starting in November. COBO (I:401) attributed the poor quality of pears in the Sierra to the fact that the ripening period came during the rainy season. Since the 1940's, a watery pear (*duquesa de Angulema*) from Germany has been rapidly propagated and is now the most highly prized fruit in the Province of Urubamba and at Hacienda Tarahuasi near Andahuaylillas in Quispicanchis. This latter estate produces about 10,000 boxes of pears a year to be sold in coastal cities.

Pyrus malus L., *manzana* or apple, does not, in general, thrive in the valley perhaps because the dormant season is not sufficiently long. The ancestral apple cultivar, perhaps dating from the colonial period, is *común*, a small inferior fruit that looks like a pippin. More modern introductions such as California delicious have not been successful and the apples of this variety which are seen in the Cuzco market are imported from Chile.

Cydonia oblonga L., *membrillero* or quince, was formerly more cultivated in the valley than now. The quince was popular in the colonial period for preserves and for a fermented drink known as *chicha de membrillo*; also, the branches of the tree were demanded for baskets (COBO: I:398). Today the above uses appear to have passed from the scene and since the fruit cannot be consumed fresh, it is now only occasionally planted.

Mespilus germanica L., *níspero* or medlar, is an Old World tree yielding a small orange-colored fruit. It probably was introduced in the sixteenth century from Spain and early became one of several rosaceous fruit trees in the Urubamba, Yucay and Calca areas. It occurs always as a lone dooryard tree and never is planted in solid orchards. The country people do not value the fruit very highly for they often do not bother to harvest the fruit. Nevertheless it is frequently sold in local and regional markets during May and June as the cheapest fruit available. The fruit is stewed because raw it is barely palatable.

Kageneckia lanceolata R. & P., *lloke*, is a shrub or small tree with extremely flexible but strong wood found in the valley above 2,700 m. Today it is rare, probably because of man's past exploitation. Many bridges over the Vilcanota River were made of braided branches of lloke (RAIMONDI, 1898:247) until the twentieth century. In his travels to Calca in 1865, RAIMONDI (1898:13) noted that 25 people were entrusted with replacing the bridge of lloke wood every 2 years. Today the wood is used for making clodcrushers. The leaves, said to contain a black dye, are no longer used for this so far as I was able to determine.

Polylepis incana H.B.K., *queñuar*, is a densely branching small tree, 3 to 8 meters high, with a twisted trunk, exfoliate bark and trifoliate leaves. It forms small copses at extremely high elevations, growing up to 4,500 m. Today queñuar is almost absent from the main valley, but it is still found in some of the more inaccessible lateral valleys which lead up to glacial cirques. The wood is white and compact, although soft and light. It makes good firewood and has undoubtedly been collected for this purpose for many centuries. It has also been used for agricultural tools and as a source of tannin.

LEGUMINOSEAE

Cassia spp. includes three closely related species of hedgerow plants (*C. tomentosa* L., *C. Hookeriana* Gill.; and *C. latepetiolata* Dom.) found in the upper valley between 2,500 m and 3,600 m. The branches are used as fuel and to make baskets, and the flowers are sometimes boiled and eaten as a famine food.

Caesalpinia spinosa (Molina) Kuntze, *tara*, is a common hedgerow tree from 2,500 m to 2,900 m. MACBRIDE (Vol. 13:Pt.3:196) indicated that it is 'commonly planted for tanning and as a source of dye' but in the Vilcanota Valley it is just as frequently seen growing uncultivated on dry slopes. Tara grows to a height of 4 to 7 meters, has yellowish flowers and bears red pods which contain, in addition to the seeds, a white powdery substance used in tanning. It is the most widely used natural tanning substance in Peru, yet there is little organized collecting of it in the valley. Formerly the wood yielded a yellow dye, but this is no longer extracted; in Yucay (GONZÀLEZ, 1833:51) much dye was prepared from the wood in the earlier part of the nineteenth century.

Inga adenophylla Pitter, *kusillo pakay*, is by far the most important of the *Inga* species in the valley. It grows from 1,900 m down valley in either a cultivated or volunteer situation. Cultivated, it is a shade tree for coffee for which it is well suited. It is evergreen and has a spreading crown of thin branches which filters but does not obstruct light. It also grows rapidly and has a deep root system which resists wind toppling. The sweet white pulp around the seeds is sometimes eaten, but it does not compare favorably with the domesticated pakay in this respect.

Inga Feuillei D.C., *pakay*, is a cultivated dooryard tree below 1,500 m. It bears long pods which contain, in addition to the six or so large black seeds, a sweet of cotton-like filament; it is this latter substance which is eaten. Pakay is well represented in the archeological record of the Coast (TOWLE, 1961: 47-8) and it probably grew in the valley in pre-Conquest times as one of the tidbit delicacies of the Incas. Today it is seen occasionally in peasant markets, but seems to be phasing out as a dooryard component because other fruit trees are now held in higher regard.

Leucaena trichoides Benth., *chamba*, is a small tree volunteering below 2,000 m. The wood is a well-regarded combustible and the leaves are used as fodder for cattle. Chamba is planted as a shade tree for coffee for which purpose it was originally brought from Central America in the nineteenth century.

Piptadenia grata (Willd.) Macb., *wilka*, is a wild tree with finely pinnate compound leaves and small teeth-like protuberances on the trunk. It occurs mainly between 1,700 m and 1,000 m. Inside the two-valved fruit are seeds which have a laxative effect when taken in small quantities or a narcotic effect in large doses. Powdered wilka seeds were snuffed up through tubes made of metatarsus of a llama (SAFFORD, 1916:558); these tubes have been uncovered at Machu Picchu from ruins of the Inca period. No evidence was found that this practice still persists. The wood of this species is considered to be the best fuel in the area that the tree grows. Consequently it has now become quite rare, except on very steep, almost inaccessible slopes.

Prosopis chilensis (Molina) Stuntz, *algarrobo*, is a common wild tree on the river banks below 1,500 m, but found sparingly up to 2,500 m. The wood is used for making charcoal and in construction and many hacienda buildings in the lower valley are of this wood. During the dry season, cattle are known to eat the pods, and in times of food shortage, people have been known to eat the sweetish edible pulp inside the pod.

Vicia faba L., *haba* or broad bean, is one of the most important crops in the valley. The broad bean is an erect plant with many leaves and the white flowers have a conspicuous black spot. The pubescent cylindrical pods have three to four flat edible seeds. Introduced from Spain in the sixteenth century (COBO: I:409), the broad bean is perhaps the only European food plant carrying the stigma of being 'Indian food'. In Spain the broad bean served primarily as livestock feed and was food for humans only when other food sources were scarce (ALONSO DE HERRERA: I:156). When the lowly Spaniard came to the New World, his social status automatically elevated, he stopped eating the broad bean. However, the Indians quickly found it better than their native tarwi, and began growing it in large quantities in areas above 3,300 m. Today it is the most widely grown leguminous seed plant in the upper valley.

Several different cultivars may have developed in the Sierra through natural and human selection to fit different environmental conditions but this would be dif-

Fig. 52. Men with eucalyptus poles threshing a pile of broad beans near Chilca, 2, 745 m.

ficult to prove. Precocious kinds are grown which complete the life cycle in 5 months; these are collectively known as *chuchín*; they have thin pods and small seeds (18-20 millimeters long). They are commonly green, but maroon, black, yellow, and red, and brown spotted kinds also grow. Chuchín broad beans are cultivated on unirrigated slopes everywhere above 2,700 m; above 3,550 m they are raised on the flats under irrigation, and they yield as high as 3,850 m. Typically, however, they are grown without irrigation and if the rains do not come during the early stages of growth, the yield is very small. Usually they are sown between October and January, and harvested from January to April or early May. In 4 months they are ready to be eaten fresh, in 5 months, to be dried. Considered 'floury', this group of broad beans is usually dried and is an essential staple food for many country people above 3,300 m.

Another group of broad beans completes its vegetative cycle in 6 months; cultivars of this group have large seeds (25-32 millimeters) with plants $1\frac{1}{2}$ meters high. The pods are thicker and wider than chuchín habas. This group of broad beans is often sown on irrigated lands in June and July and is harvested in December or January in the valley between 3,000 m and 2,600 m where they survive the generally mild frosts of the low-sun period. More commonly, they are put into the ground in late September or the early days of October and usually are not fertilized. Below 3,000 m broad beans are not grown in individual fields but are mixed with maize. As with chuchín beans, a wide color variation exists, but the most common is a light green variety. The peasants consider these large broad beans to be watery and prefer them in fresh form rather than dried.

Except on steep slopes, broad bean fields are prepared with the plow and oxen. In Canchis, habas are sown in rows about 50 centimeters apart with one to three seeds per hole. The plants are harvested with a sickle. Small bundles are tied together and made into sheaves. These sheaves are left to dry 6 or 7 weeks, and many Indians believe that the beans can still draw nutrients from the cut stalks in the sheaves. When thoroughly dry, the beans are threshed on a threshing floor by animals, or beaten with poles if the crop is small (Fig. 52); later they are winnowed. The dried stalks are fed to cattle.

Broad beans are appreciated as a crop because of their high yields. An agronomist estimates the yield of the dried kernel to be 1,500 kg/ha as compared to 1,250 kg/ha for dried maize, and 900 kg/ha for dried kidney bean (*poroto*). Green habas yield 7,000 kg/ha. Yields vary because of disease and insect pests which appear to increase their damage as one moves down the valley. Thus at Ollantaytambo, broad beans are much less common than in Canchis and Quispicanchis. Broad beans are consumed fresh and dried; a very popular peasant dish in the valley is boiled green habas (*puspo*). Dried broad beans are made into flour and eaten as mush, or they are fried in oil and salted, and eaten as a confection.

Pisum sativum L., *arveja* or pea, appears to be rapidly gaining importance in the peasant agriculture. It is not certain when and from where the pea was introduced because COBO or other chroniclers do not mention it in their works. Peas are grown in the valley from about 3,800 m to 2,800 m but always in small amounts. The seeds of a white-flowering cultivar are dried and ground into flour to be used in stews and soups; while the pink and purple-flowering cultivar is boiled, toasted or eaten fresh when still in the maturation stage. To be dried, the pea vines are stretched out their full length in the sun and left until no moisture remains. When completely dry, the whole plant is cut off and carried to the threshing area. Rarely does the amount harvested justify threshing by animals; more commonly the plants are flailed with sticks. The dried vines are valued as fodder for livestock.

Cicer arietinum L., *garbanzo* or chick pea, is a crop clearly introduced in the sixteenth century from Spain (GARCILASO DE LA VEGA, 1960:447). Although formerly it may have been important, it is today very uncommon in the valley and does not form an element in the diet of the people as it does in other valleys of the southern Sierra. Part of its absence is related to the lack of a suitable ecological niche here, for chick pea thrives best where rainfall is low and temperatures are high.

Arachis hypogaea L., *inchis* or *maní* or peanut, is grown below 1,600 m in the Vilcanota Valley and probably has been since at least Inca times. Formerly it was a commercial product of several estates but today little is grown for sale outside the lower valley. Peasants may grow a few rows of peanuts on their chacras, but they do not depend on them. It is only partially integrated into the peasant diet and economy perhaps because people prefer to grow a crop that is also known in the Highlands, such as kidney bean. The peanut is boiled in salt water, toasted, made into a flour for mush and occasionally made into chicha de maní. While the peanut is one of the world's important sources of edible oil, to my knowledge, it has never been used for this in the valley.

Trifolium amabile H.B.K., *chikmu*, a native plant, grows in weedy places in the upper valley. It is collected as a potherb during periods of famine and at other times is fed to cattle.

Medicago sativa L., alfalfa, is the major forage crop planted in the valley, and has grown there for approximately 400 years. It was introduced into Lima from Spain in 1535 (COBO: I:410) whence it spread into the Highlands. It first became important as a crop when it was cultivated to feed the mules and horses of the Cuzco-Potosí trade. Agronomists assert (SWANSON, 1956:1) that a special mountain type of alfalfa has evolved in the Sierra which is hollow-stemmed, grows best in cool weather and lives 10 to 20 years or more. Alfalfa is grown in the valley today as high as 3,595 m by farmers who feed it to their livestock or who grow it as a cash crop to sell in small bundles at local markets. Before the adoption of a hardy variety, alfalfa was cut and dried because otherwise the frosts would kill it (RAIMONDI, 1898b:242). Today, however, alfalfa grows year round, is always cut and used green, no longer made into hay, and the plant is kept about seven years before the land is reseeded with fresh stock.

Melilotus alba (Med) Desr and *M. indica* (L.) All are two Old World weeds that are pulled out of crop fields and fed to cattle in the upper valley.

Indigofera suffruticosa Mill., *añil* or indigo, yields from the leaves a blue dye which was used for textiles manufactured in pre-Columbian Peru (YACOVLEFF & HERRERA, 1934:268). Indigo has been cultivated in the lower Vilcanota Valley, but it is uncertain if it was prior to the Conquest. Evidence of its cultivation and use in the Inca period is absent, and it may have been brought to the valley by the Jesuits who were eager to exploit the commercial possibilities of any crop amenable to plantation-type agriculture. The earliest reference to indigo was made by COBO (I:216) who in 1610 mentioned *añil* as one of the crop plants of the Cuzco Yungas. It may have been brought here about the same time that sugar cane was introduced, and it grows under the same environmental conditions. It was extensively cultivated in the lower valley during the middle of the nineteenth century (RAIMONDI, 1898:131; GONZÁLEZ, 1833:301). Peasants probably never grew or marketed it because indigo required an elaborate and costly processing which was beyond the means of the natives. In the late nineteenth century, indigo cultivation ceased in the valley because a cheaper and better synthetic blue dyestuff was perfected in Europe. Today in the valley, indigo is found only as an uncommon weed; this is the only visible testimony to its past importance.

Astragalus garboncillo Cav., *huska*, is a wild plant found at elevations above 3,400 m. The leaves have use as a poultice, but the medicinal value of this plant is far outweighed by its deadly poisonous effects on cattle and sheep who eat it.

Psoralea pubescens Pers., *wallwa* is a malodorous wild shrub valued for its blue flowers which are used as a purge.

Lupinus mutabilis Sweet, *tarwi*, is a native domesticate grown for its edible seeds. It is an erect herbaceous annual with blue flowers born on a racime. The height depends greatly on fertilization; if the land is fertilized, the plant will grow to about $1\frac{3}{4}$ meter, if not, only to a meter. The slightly hairy pods are 8 centimeters long and 16 millimeters wide, and each plant has a total of about 25 pods with an average of five seeds in each pod. Field observations indicate that two kinds are found in the valley: narrow pods with white seeds and wide pods with white seeds with black spots. Whatever the kind, tarwi seeds are flatter, smaller and more rectangular than the seeds of the common kidney bean.

Tarwi very probably was domesticated in Peru, if only because several of the wild species which resemble tarwi are found in this part of the Central Andes. The time of domestication is not yet known even generally, for no archeological evidence has been found. In pre-Conquest times (before the introduction of small grains and the broad bean), tarwi was a major source of vegetable protein. The first historical reference to tarwi was in 1539 when VALVERDE (HERRERA, 1941:4) in a letter to the King suggested a tax on this crop to create revenues. The Spaniards first called it *altramuz* after *Lupinus albus* which was cultivated in Spain at that time, also for its edible seeds.

In the Vilcanota Valley, tarwi is clearly a relict crop. Whereas around Lake Titicaca it is still a conspicuous agricultural component, in the valley it is now relatively uncommon. In 1972 (ANONYMOUS 1792:13), *chochos* (as the Spaniards later called tarwi) was still an important crop in the Sicuani area perhaps on a par with broad beans, wheat and barley. Today tarwi is grown but sparingly from 3,850 m to 2,600 m, with its greatest agronomic emphasis between Tinta and Combapata. Solid fields of tarwi are rare and usually it is planted in a corner of a maize field or on the periphery of another field to act as a barrier (GADE, 1969b).

The great amount of work involved in processing the tarwi for consumption may partly explain its decline and its replacement by broad beans which resist equally well the low temperatures but which do not require this processing. Tarwi seeds must be boiled and then soaked in running water for 3 or 4 days to rid them of a bitter principle. They are usually served in a fresh salad with capsicum pepper, onions and herbs, or eaten as a confection (Fig. 53). Occasionally dried tarwi is for sale in the Sunday market at Sicuani as one of the lowest-priced foods at this fair.

Some estate farmers grow tarwi not as food but as a cover crop because of its nitrogen-fixing properties. One wonders if the ancient inhabitants of this land knew in an empirical way the value of the legume for enriching nitrogen in the soil.

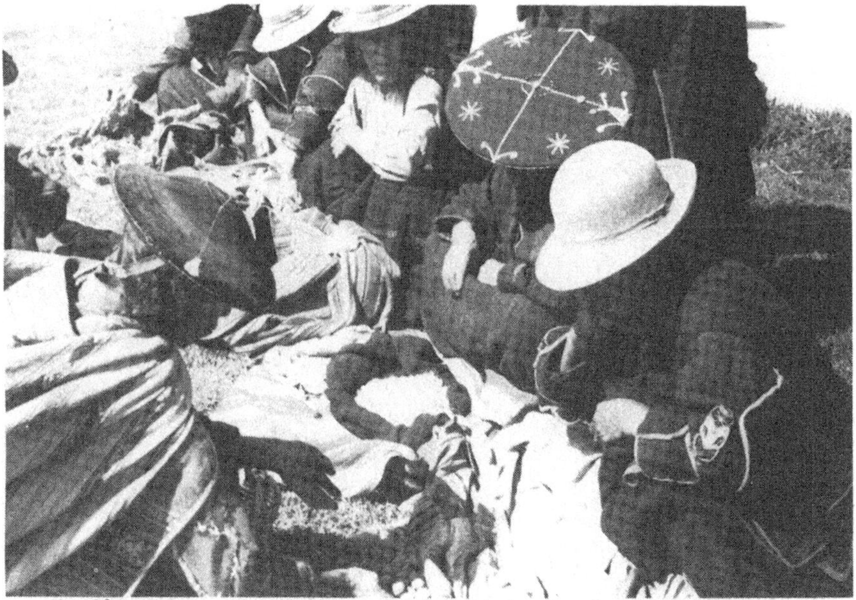

Fig. 53. Indian women selling tarwi, already washed and cooked, in the Tinta market.

Spartium junceum L., *retama*, is a broom-like shrub with nearly leafless branches and papilionaceous yellow flowers that bloom throughout the year. COBO (I:405) noted that retama was brought to Arequipa from Spain in 1850; thus this plant has been in the valley for at least 300 years. Today retama covers large areas in the valley between 3,100 m and 2,600 m but can grow from 3,350 to 2,400 m (Fig. 43). This plant has a very weedy nature and occupies several habitats in which it does not have to compete with the native flora. Areas of former landslides, roadsides and trailsides and fringes of eucalyptus groves are all favorable habitats for this plant, as are sandbars and riverbanks. Retama has become an important thatch material from 2,950 m to 2,500 m and is considered to be more water-resistant than grass thatch (Fig. 54). Maize ears are dried in a 'bed' of retama branches which protects the maize from ground moisture. Homemade brooms are also made from the branches and sheep and goats will browse on the plant. It also plays a role in folk medicine: flowers and seeds are used as purgatives, leaves for treating kidney stones and the root for liver ailments.

Batesia floribunda, wayruro, a tropical tree that yields brillant red oval, poisonous seeds called wayruros, grows in the Eastern Lowlands but probably not in the lower valley. In Inca times the seeds were used in witchcraft (COBO: I:272). Today in peasant herb markets those seeds having a black spot are known as 'casados', whereas solid red seeds are called 'solteros,' indicating their use as love charms. They are also strung to make necklaces which are sold in tourist markets.

Fig. 54. A cholo at Urubamba carrying a load of retama branches to be used as thatch.

Dolichos Lablab L., *andreja* or hyacinth bean, is grown as an annual crop for its edible, although somewhat bitter seeds at elevations below 2,500 m (Fig. 40). The hyacinth bean is presumably of African origin and was perhaps brought to the New World and to Peru by slaves. This plant may have been introduced into the valley around the end of the sixteenth century when slaves were brought to the Quillibamba area to work in the sugar cane fields. Today it is a distinctly minor crop and is not seen in peasant markets.

Pachyrrhizus tuberosus Spr., *jíquima* or yam bean, is a leguminous plant with poisonous seeds and edible tuberous roots. It was found in the pre-Inca period on the Peruvian coast (TOWLE, 1961:52) and was a widespread cultigen in the New World before the arrival of the Spaniards, COBO (I:69) mentioned 'giquimas' as a product of the Cuzco warm lands. Its ecological position is below 1,500 m in the valley but in the area under study, the crop may be extinct. Inquiries here suggest that since it is not regarded as a nutrious or tasty food, there is no incentive to grow it. Farther down the valley, however, in Machiguenga territory, jíquima is still part of the crop inventory.

Krameria trianda R. & P., *pacha lloke* or rhatany, a branching shrub, grows on hillsides that border the Vilcanota River above 2,700 m but below 3,400 m. The cortex of the root contains a powerful astringent (recognized by the U.S. Pharmacoepia) and is highly valued as such in folk medicine. Attempts in Cuzco to commercialize the root abroad failed.

Phaseolus vulgaris L., *poroto* or the common New World kidney bean, is found in the valley but does not form a major element in the peasant diet as it does in Mexico. Its large range extends from about 3,300 m to below the study area, but within this environmental range (Fig. 40) the kidney bean is always grown in fields with other crops. In the upper valley, if harvested in sufficient quantity, the beans are threshed by animals in the same manner as the broad bean. If only small amounts are grown, they are flailed with a stick.

GARCILASO DE LA VEGA (1960:371) noted four kinds of *'frisoles'* (*sic*) which were grown in the Cuzco region in the sixteenth century, but it is not clear if he meant only. *P. vulgaris* or if he included other species as well. The two most common cultivars today are *chekche* and *blanco*; the former is cream-colored with brown spots, the latter, white. Between Pachar and Urubamba, the chekche bean is a commercial crop for certain peasants who roast the beans and sell them at regional fairs and to travellers as a confectionary snack. Below 1,500 m a small white-seeded cultivar known as *panamito* is preferred and can be harvested within 3 months after being planted. Another cultivar of the lower elevations is *canario*, a small yellow-seeded bean, said to have come from the Andahuaylas region (180 kilometers north of the valley) which may be a major varietal center for the kidney bean in the Andes.

One of the factors that has limited cultivation of kidney beans is the long time required to cook them when dried. At high elevations, where fuel is at a premium, this is a distinct disadvantage. Also, the cool nights above 2,400 m are said to hinder their growth. Peasants complain of the high cost of seed and this initial expense may prevent many of them from cultivating this plant.

Phaseolus lunatus L., *pallar* or Lima bean, is cultivated on a very small scale below 2,800 m. Above this elevation, the Lima bean does not grow satisfactorily because of the climate, and it is used primarily as a traditonal item in ritual offerings to Mother Earth.

Tamarindus indica L., *tamarino* or tamarind, is a large, leafy tree that may have been introduced to the lower valley by the Jesuits. The fruit is a pod, and contains dark red pulp with an agreeable tart flavor. It has been used as a medicinal and, more importantly, as a refreshment.

Erythrina falcata Benth., *pisonay*, is a tall native tree which, in its natural setting, is practically the only tree along the Vilcanota River between 2,300 m and 2,100 m. Covered with epiphytes, these trees lose most of their leaves in the dry season which is also the period during which they are covered with brilliant red-orange blossoms. Outside of its natural range, pisonay also is a planted species. Conspicuous individuals are seen in the upper valley at Urubamba, Yucay, Urquillos, Huayoccari, Calca, Taray, Pisac and Andahuaylillas. In his travels to Calca in 1865, RAIMONDI (1898:13) noted the huge old pisonay trees around the church – the same ones found there today. In 1904 LORENA (1932) noted the large pisonay trees at Yucay, 26 meters high and 3 meters in diameter at the base, which he estimated to be more than 400 years old; shortly afterwards wind knocked them down. The wood of pisonay is very poor for construction or even as a firewood, and the tree's value has always been mainly an ornamental one.

Tephrosia toxicaria (Sw.) Pers., *kumu* or barbasco, is a wild or semi-cultivated shrub of the lower valley which, when dried and powdered, is used as a fish stupifier. Earthen dams are made to form small pools in the river for this purpose. However, most fish that are caught in the Vilcanota study area are removed by hook and line; whether this has always been the case is not certain.

OXALIDACEAE

Oxalis tuberosa Mol., *oca*, can be distinguished from other tuber crops by the scale leaves which border the tuber eyes. The tuber is cylindrical and although it may be any one of a number of colors, the common cultivar in the valley is pink. Oca is grown to some extent in southern Canchis, but is is scarce in the rest of the valley. (Fig. 39). Rarely is oca planted in solid fields; usually one sees two or three rows of it alternated with ulluca and/or añu. Oca tubers contain a large quantity of starch and they are appreciated for this quality. Like añu, it is placed in the sunshine for approximately 8 days to 'develop' the sugar content, but also to rid the tuber of the unpleasant biting sensation caused by the calcium oxalate crystals. Oca may be roasted, boiled or dried. The bitterest kinds of oca are dehydrated to make *k'aya*, now very rare in the Vilcanota Valley but made on the Altiplano.

TROPAEOLACEAE

Tropaeolum tuberosum R. & P., *añu* or *isañu*, is readily recognized as a nasturtium by its peltate leaves and bright red-orange flowers with a spur. The deep-set eyes of the añu tuber are without scale leaves. The size, shape and color of this domesticated tuber crop vary less than ullucu or oca. In the valley *kello añu*, a yellow cultivar, is by far the kind most widely grown. Large quantities of añu are grown in the community of Occobamba (3,850 m) near the upper limit of crop cultivation (Fig. 55). Although the crop is very hardy and thrives in this microthermal environment, its resistance to plant disease goes further in explaining the great emphasis on it. Although often regarded as the most unpalatable of the tuber crops, the people here have obviously acquired a taste for añu.

Elsewhere in the valley, rows of añu are occasionally seen from 4,000 m down to 3,300 m (Fig. 39); below this elevation, añu is grown only on the uppermost slopes of the valley. In Canchis, the tuber is known as isañu, an Aymara term; this suggests that the plant was brought from the Altiplano. More than any other tuber, añu is restricted to home consumption and is not an important trade item. Only

Fig. 55. The añu harvest at Occobamba in southern Canchis (3,890 m). Small piles are made which are later added to the large pile in the foreground.

occasionally does one see it in markets in the valley. In the Yucay area, the country people do not like añu, but will readily acquire other tubers (oca and ullucu). Its reputation as an anti-aphrodisiac, its unpleasant taste and its nutritional inferiority explain its very localized importance in climatically difficult areas.

LINACEAE

Linum usitatissimum L., *lino* or flax, was a sixteenth century introduction into the New World and may have been brought to the Cuzco region in that period (GARCILASO 1960:450). In early colonial times (PUENTE Y OLLEA, 1900:415), the seed of the flax was valued for its oil which was used to make artist oil paints for the innumerable canvases turned out by the prolific mestizo artists of the seventeenth century 'Cuzco school.'

In the nineteenth century, flax was again cultivated for the oil in its seed in the Province of Calca. The oil was used to prepare emolients. In this century several estates, among them Hacienda Chillca and Hacienda Primavera (both about 2,700 m. in the valley), tried to grow flax for fiber. The flax reportedly grew well, but there was no market for the fiber and its cultivation was discontinued. Fiber flax is a crop which would fit well into the agricultural scheme of the valley because it can withstand the low average temperatures of the area, and also because adequate supplies of cheap, patient labor are available to ret the fibers. Today in the Vilcanota Valley flax is a medicinal plant usually grown around dooryards. The toasted seeds are crushed and used in an infusion to cure stomach pains and fever.

ERYTHROXYLACEAE

Coca

Erythroxylon Coca Lam., *coca*, is the species of coca grown in the Vilcanota Valley and in most of Peru.[25] No real cultivars exist, but many minor variations in leaf size and shape can be found; one such variation recognized by the coca growers is *chimpe machu* a coca leaf with a very pronounced apex. Coca is cultivated in Peru and Bolivia and, on a very small scale, in Colombia, Venezuela and Argentina. Apparently it is no longer grown in Ecuador.

Over 900,000 Indians in Peru masticate the coca leaf. Perhaps 80 percent of the chewers are adult men; 20 percent are adult women. The daily intake of the habitual chewer varies from 10 to 100 grams, the largest intake being by Indians living at very high altitudes. Whether or not coca is a pernicious drug that should be stamped out is a matter of much controversy. Many observors contend that it numbs the chewers into a sub-human state, while others believe that this anaesthetic effect actually benefits him because it wards off hunger pangs caused by a deficient diet. There is also little agreement about the toxic effects of coca and to what extent the natural cocaine contained in the leaf is habit-forming.

Coca is used by stuffing the dried leaf into the mouth and chewing it in quid form. A bit of *llipta*, a putty made of quinoa ashes, is taken to help extract the alkaloid from the leaf. Besides being chewed, coca has sacred and magical usage in religious offerings and as a folk remedy it is taken for a wide range of maladies. It is used either in infusion (for example, to relieve asthma and indigestion) or applied as a poultice (for example to ease rheumatic pains). Some coca grown in Peru is used to manufacture cocaine, either legally in the government factory in Lima, or clandestinely in small establishments throughout the country.

Historical Geography of Coca

Coca may have been first cultivated by forest Indians somewhere in the western part of the Amazon Basin. Today a few tribes in Brazil grow coca and almost all forest groups along the eastern base of the Andes cultivate it. Before the highland

25 *E. novogratense*, the coca grown under irrigation in the upper parts of certain coastal valleys in the Department of La Libertad, is more tolerant of high temperatures than is *E.Coca*. Of these two cultivated species, the former has a smaller and more pointed leaf and is less rich in crystallizable cocaine.

people themselves grew coca, they may have acquired it by trade from the forest tribes. When it was first used in the Highlands is unknown. On the west coast, the earliest archeological evidence dates from about 500 A.D. (TOWLE,,1961:140). The areas in which coca was most intensively grown were nearest the city of Cuzco. Apparently only the privileged classes in Inca times were permitted to use coca and its consumption was limited.

After the Conquest, coca became universally used by the indigenous population. Long llama or mule caravans, each carrying 2,000 to 3,000 baskets filled with dried coca leaves, transported huge quantities of coca from the lower eastern valleys in the Department of Cuzco to the mining settlements in Potosí to supply the Indians who worked there (GAGLIANO, 1963). Such caravans treked back and forth between Cuzco and Potosí many times annually until mining declined in the seventeenth century. Production fluctuated greatly because Indian coca laborers were scarce since they were apt to contract leishmaniasis. Exploitation of these laborers by the Spaniards led to legislation regarding coca beginning in 1571. An order by the Viceroy Toledo (ZIMMERMAN, 1938:139) to limit the planting of coca was, however, rescinded because of the economic pressure of over 2,000 Spaniards who had plantations and who sold the leaves to more than 300,000 Indians. In 1572, new laws stated that no Indians were to be forced to work in the coca plantations and, furthermore, that the Indians were not required to pay tribute in coca leaves. All bridges and roads in the areas of cultivation (ZIMMERMAN, 1938:140) were to be repaired by the owners of the coca plantations and in 1576, a 5 percent tax was placed on all coca gathered. These laws were ignored after about 1650 when coca from Cuzco was no longer traded to Potosí.

In 1803, a coca revenue office (*alcabala de coca*) was again established to collect a tax on all coca shipped out of the producing areas. Much of this revenue was used (RAIMONDI, 1898:136) to repair roads and construct bridges in the Vilcanota Valley in the nineteenth century. It would appear that the valley did not become really important for coca growing until the nineteenth century, although the crop was grown to some extent in Inca and colonial times. Because there are no Inca ruins in the lower part of the valley, it is difficult to assert that the valley was an important coca growing area in the pre-Conquest period. Until well into the twentieth century, the area around Chaullay (1,200 m) was the center of production, for not only was this the most accessible part of the valley, but it also had optimum environmental conditions. As transportation has improved, coca has migrated down the valley. But in both pre-Conquest and early colonial periods, the eastern slopes of the Province of Paucartambo (Fig. 15) – then known as *Los Andes* – was the most important coca growing area in southern Peru, for it was the tropical area closest and most accessible to the city of Cuzco.

Coca Ecology

Coca yields well when average annual temperatures are between 18° and 25° C. Temperatures definitely retard growth if they average below 10° or above 35°C.

Because of temperature limitations, coca is grown in the Vilcanota Valley between 1,920 m and 730 m (and lower, outside the study area). At these lower elevations, the coca plant transpires more water than it can bring up through the root system and consequently coca is found here only in very small plantings. Above the upper limit, the low temperatures as well as cloudiness, which affects photosynthesis, retard the development of the leaf and frost darkens the leaf and imparts to it a bitter flavor. The highest yielding and best quality coca comes from between 1,300 m and 1,000 m altitude, and coca grown within this range brings higher prices than those paid for coca grown below 1,000 m. Below 1,500 m coca that is grown on the upper slopes (*altura*) is harder, thicker, sweeter and contains more cocaine and sells for 3-4 soles more per kilogram than that grown on the river bottom (*playa*); the latter sites yield coca which is more paper-like and contains less cocaine. From San Pablo (1,420 m) to Quillabamba (1,080 m), most of the upper slopes (100-300 meters above the valley bottom) are devoted to coca.

In the Vilcanota Valley the lands which are now in coca were formerly used for other agricultural purposes. This is unlike the other major Peruvian coca producing region. The Department of Huánuco in central Peru, where coca lands have been made from cleared brush or forest land. In the valley manioc is regularly planted between every third row of coca plants and provides food until the coca shrub is mature enough to be harvested. When coca becomes old (after 15-20 years) and must be replaced, the young plants are gradually intercalated between the old rows.

Coca is grown on either irrigated or unirrigated lands. Irrigated coca is always on the flattish bottoms of the valley. Because these plantings are not affected by the dry season they may be harvested four times a year. Unirrigated (*temporal*) coca plantings are grown either along a contour or in up and down rows on slopes, often very steep ones, and only three crops are harvested per year because there is little leaf growth during the dry season, July to October (Fig. 56). The coca lands are hardly ever fertilized, nor is the soil naturally rich. Most of the coca is grown on soil weathered from slate and extremely poor in humus content.

Coca Cultivation

The methods of cultivating coca are highly traditionalized, and what was observed well over a century ago by GRANDIDIER (1861: 109-113), has not substantially changed. Reproduction is always by seed, never by cuttings: vegetative reproduction has worked experimentally but it is not a traditional method of the peasants, and they have not yet accepted it. The fruit of coca is a drupe which turns brown when fully ripe. The seeds are collected, a task separate from the collection of the leaves, when the fruit turns yellow just before it is fully ripe. The ovoid fruit (*mokllo*), the size of a corn kernel, is put in a basket and placed in the shade for 5 days or so to permit the pulp to soften; the pulp is then removed by washing. After being dried in the sun, the seeds are stored in a dry place.
Coca is started in a seed bed for a variable length of time and then transplanted.

Fig. 56. On the left, coca growing at its highest elevation in the valley at 1,950 m. On the right, sugar cane.

The seed bed (*wambal* or *almácigo*), usually 4 meters by 12 meters, is made between December and March. The seeds germinate in about 24 days. When the plant reaches the four-leaf stage (known as *chiquito*), a latticed covering is erected 1 meter from the ground to shade the tender plants. After a year in the seed bed, the plants (*wata-planta*) are 30-40 centimeters high, and they are then transplanted in January or February, during the rainy season. Two or three plants are placed in holes along furrows that are made 90 centimeter apart. Some coca (*chuco*) is grown without the benefit of artificial shade in the seed bed and transplanted to the field six, rather than twelve, months after planted.

From time to time the fields are weeded and earth is banked up around the roots (*aporque*), a procedure similar to that used in native maize cultivation. As with maize, the weeding and banking are done with the lampa or kuti. Mulching, a European practice, is not done. Artificial shade is needed for the plant until it is 2 years old; this is provided by manioc which is planted every third row, 6 months before the young coca shoots are transplanted from the seed bed. Coca also must be pruned (*cushupa*). This is done in September and October when the plant is still dormant. The tops of old shrubs which are attacked by witches broom disease are cut off with a special knife (*cushupeador*).

The coca plant is attacked by insects and fungi. Weevils and crickets damage the shrub some, but the leaf-cutting ant (*kuki*) is by far the worst pest. These ants do most of their damage at night, and they are especially plentiful in November and

December when they destroy many seed beds and producing shrubs. The most effective control measure is to burn them with kerosene when they 'march.' Two fungus diseases which cause considerable damage are witches broom (*k'upa*), produced by an agarical fungus, and *ayawaiko*, caused by a *Fusarium* fungus which makes the root and the leaves turn yellow.

When 3 years old, coca plants begin producing small harvests of leaves. The first harvest (*plantilla*) yields small choice leaves which are often festively distributed to friends of the owners. The leaves from the second harvest (*linli*) are somewhat larger, but the big harvest (*waranchi*) is usually the third year when the plant is 6 years old. A coca shrub may have as many as 50 or more years of productive life, but when the plant is 20 years old, the leaves are small and of inferior quality. Some plantings are kept only 10 years.

The leaves are picked, one by one, a task which requires great patience and is usually done by women. A good coca picker (*palladora*) can pick 40-45 kilograms of leaves in one day. Ordinarily she works down a slope rather than along the contour. The collected leaves (*mato*) are taken to a shed (*matowasi*). There the leaves are placed on wooden platforms for 2 or 3 days and must be constantly turned to prevent fermentation which would lower the quality. From the platforms they are taken to the drying yard (*matokancha*), a floor of inlaid slate or other stones where the coca is spread out evenly, about $2\frac{1}{2}$ centimeters deep, to ensure uniform drying by the sun. No drying is done by machinery in the valley. When the drying leaves begin to curl, they are turned over by crudely made brooms (*hukucha chupa*). If coca is not turned, the leaf will become so superficially dessicated that it will crumble. Once the leaves are satisfactorily dried to a parchment-like texture, the impurities are sifted from the coca in a wire-mesh screen which removes coca seeds, stones, dust, twigs and leaf particles. Only then is the coca ready for packing.

Coca Production

The quality of coca is more a function of how it has been processed than of its growing environment or of botanical differences. First class coca (*k'acha*) is that which has been dried in only one day. It loses approximately 60 percent or so of its weight in the drying process; from 100 kilograms of fresh leaves, about 40 kilograms of dried leaf are obtained. K'acha coca is uniformly green in color and contains a higher percentage of alkaloids than other classes because it has lost less cocaine by evaporation. Extra young leaves in this class generally fetch the highest price. Second-class coca (*pisada*) sells for about 12 percent less than k'acha. It is coca which did not dry rapidly because of rain or lack of sunshine. To be dried, this coca is trampled upon or flailed with sticks. Often large stones are rolled over the leaves before they are spread out to dry. These procedures make the coca leaves turn very dark green or blackish-brown and, despite these operations, this coca still has more moisture and a 25 percent greater volume than k'acha has. The third class of coca is *coca lavada*, a brown undried coca, and often is one fourth by weight leaf parts. Fourth class coca is fermented leaves; it is useu in folk medicine and magic, but not in commerce.

Coca is overwhelmingly produced by peasants who grow it on land rented from an estate.[26] Until the twentieth century, coca was part of hacienda production, but the increased cost of labor and great fluctuations in price forced the hacendados to abandon coca cultivation for other crops. Today only about 6 percent of the coca is produced by the estates themselves, and almost all of this small percentage is grown on irrigated lands near the river.[27]

Coca is packed for market in bundles of one arroba (=10.4 kilograms). Straw is packed around the leaves and each bundle is wrapped in banana trunk fibers. In 1963, an arroba of first-class coca was worth S/180. at its place of production. To this price was added S/40. for government tax and S/2.50 for transportation. Thus coca in the city of Cuzco was sold for S/223. for 1 arroba (about US$0.36/1b.). Ninety five percent of the coca from the lower valley is used for mastication, and probably 1 percent for medicinal purposes and 4 percent for export.

A government agency, the Estanco de la Coca, controls the sowing, cultivation and harvest of coca as well as its distribution and exportation. The agency also collects the government taxes which are imposed on coca production. All commercial coca can be marketed only through stores which are authorized by the government to sell this commodity. In the city of Cuzco, there are 428 such outlets, in Quillabamba 294, and in Sicuani, 68. These shops in turn sell coca to cholos in small villages who resell it to Indians in the local peasant markets. The early importance of Sicuani was tied to its position as a major coca market, redistributing the leaf to a huge area.

For several years the United Nations has urged the Andean countries to eliminate coca growing as the significant step in supressing coca chewing and illegal international narcotic trade. In 1961 Peru ratified the International Drug Convention and a 1965 Peruvian law limited the cultivation of coca to certain well-established regions. No new seed beds are allowed (theoretically), but this law is not conscientiously enforced. In fact, any rise in price is followed by an increase in

26 In 1963, declarations by districts were as follows (figures from the Estanco de la Coca):

district	production	elevational ranges
Urubamba	8,915 kg.	2,000 to 1,600 m
Santa Teresa	247,663 kg.	1,600 to 1,300 m
Maranura	1,124,027 kg.	1,300 to 1,100 m
Santa Ana	1,197,639 kg.	1,000 to 980 m
Echarati	539,321 kg.	980 to 730 m

27 At Hacienda Mandor, 37 hectares of coca were managed by the estate, while 212 hectares were cultivated by the peasants themselves. At Hacienda Chinche, it was 24 hectares of hacienda production, and 248 hectares of peasant production. Hacienda Santa Rosa (now extinct) had the largest production (702 hectares), virtually all of it produced by peasants as one of their own cash crops.

coca cultivation, clandestine or open. The elimination of coca growing would impose a hardship on most of the 50,000 peasants who grow and sell the leaf. Furthermore, because of tax revenues on it, the government is not anxious to extirpate coca; in 1963, for example, La Convención contributed S/14,560,000 (about $560,000) to the national treasury from coca. Until an equally satisfactory and lucrative commercial product is found for the lower Vilcanota Valley, its 20,000 hectares of land in coca, representing about 80 million coca plants, will probably remain, 'legal' restrictions notwithstanding.

RUTACEAE

Citrus Aurantium L., *naranja agria* or bitter orange, may have been brought to the valley in the sixteenth century by the Jesuits who used it as the grafting stock for the sweet orange. The astringent fruit, not edible when fresh, may have been used in the lower valley as it was in Spain, in preserves and as feed for pigs. Today the bitter orange grows up to an elevation of 2,200 m in the valley but it is not highly valued and is seldom seen.

Citrus sinensis (L.) Osbeck, *naranja* or sweet orange, may have been brought with the bitter orange to the valley in the sixteenth century. CIEZA DE LEÓN (1883:339) stated (for 1551) that the Spaniards planted oranges in Cuzco but he did not specify which kind. Today the sweet orange is by far the most common and important kind of citrus fruit grown. It grows below about 2,300 m and the trees increase significantly in number below 1,700 m where almost every peasant chacra has five or six trees, either around the hut or on the margins of a field. Most of the oranges seem to be descendents of an old introduction: small, thick skinned, many seeds and not particularly sweet. In 1950 the Washington Navel orange (called *Huando* in Peru) was introduced into the valley but its adoption remains limited and production is all for local use. But large quantities of the ripe fruit of the common orange are taken to Cuzco. In smaller towns above 3,000 m the fruit is purchased not so much to be eaten for the usual reasons, but to be used as a home remedy for a wide range of ailments.

Citrus Limon (L.) Borm. var. *limetta, lima* or sweet lemon, is almost as common as the sweet orange in parts of the lower valley. Like all citrus, it is native to the Old World and was brought to the valley in colonial times (COBO: I:401). It does not have nearly the commercial value of the sweet orange, but many country people prefer it as a fresh fruit. It is juicier than the sweet orange, with a sweetish liquid but very bitter tissue which is discarded. The sweet lemon is the hardiest of the citrus trees grown in the valley and occurs as far up the valley as Huayoccari, 2,945 m, however, only below 2,000 m, is it an important horticultural component.

Citrus aurantifolia (Christm.) Swingle, *limón sútil* or lime, has a small greenish-yellow fruit with a very thin peel. This fruit is important as a seasoning and to cut the greasiness of many fried foods. The juice of limes, together with those of sweet lemons and oranges, are used in folk medicine to cure such diverse maladies as insomnia, tuberculosis, leprosy, rabies and alcoholism (MOSCOSO, 1963:62-3). Limes commonly volunteer on trailsides and river banks, far from any human habitation. The large number of volunteer lime trees below 1,000 m in the lower valley was observed by RAIMONDI in his 1858 visit (RAIMONDI, 1898:41); he conjectured that this tree spread from cultivated limes on Jesuit estates.

Citrus medica L., *cidra* or citron, has probably been in the valley since the sixteenth century (COBO: 1:400). Today below 1,500 m it is sparingly grown for the rind of the fruit. Citron rind is frequently seen sliced and dried in peasant markets and is used in home remedy infusions.

Citrus reticulata Blanco, *mandarina* or tangerine, is regarded as a fairly recent introduction into the lower valley where it was supposedly introduced by an hacendado in the early twentieth century. Tangerine production has been commercialized and several estates in the lower valley maintain groves of this fruit. It has become a favorite fruit in Cuzco.

Citrus paradisi Macf., *pomelo* or grapefruit, was called *toronja* by COBO (I:401). The grapefruit that we know today has existed only since 1814 when it was supposed to have originated as a mutation in the West Indies. COBO's toronja may have been the related shaddock (*C.grandis*).The grapefruit, fairly rare in the valley, is somewhat of a curiosity grown by a few hacendados for their own use, and it is almost never seen in peasant markets.

Ruta graveolens L., *ruda*, is an Old World perennial brought to Peru in the first years of the Conquest (COBO: I:420). This herbaceous plant is sometimes cultivated but more often grows as a ruderal around peasant dwellings. The strong smelling leaves are used in infusion to aid digestion, alleviate cramps and intestinal gas, and to minimize the dire effects of 'bad wind' (*mal aire*).

MELIACEAE

Swietenia mahogani Jacq., *caoba* or mahogany, is a large tree with a hard, compact wood used in cabinet making. Its ecological position lies below 1,100 m in the valley but it is now rare in the study area and the wood must be gotten from farther down valley.

Cedrela Herrerae Harms, *cedro*, a tree of the ceja de la montaña has a highly valued wood. In the colonial period (ANONYMOUS, 1963:32-66), picture frames, retablos, pulpits and effigies in churches of Cuzco and other towns of the Department were made of *Cedrela* wood. Today it is used in cabinet making, but wild specimens are now rare in the valley. Outside of its native range, it has been planted as an ornamental at higher elevations. Many plazas in towns in the elevational range of 3,400 m to 2,800 m have large shady cedro trees.

MALPHIGIACEAE

Bunchosia armeniaca (Cav.) Rich. apud Juss., *ciruelo*, is a small tree that bears an oval red fruit with an acid flavor and a large seed. It was grown in the Inca period (TOWLE, 1961:61). The Dominican fathers brought it from Lima several decades ago to their mission farm near Quillabamba. However, the tree has not moved much from this dispersal center; the country people generally reject it because they do not like its flavor.

EUPHORBIACEAE

Euphorbia penicillata Millsp., *wachanka*, is an annual herbaceous plant known to be a purge. The tuberous roots, not the seeds, contain the purgative property which was well known in Inca times (GARCILASO DE LA VEGA, 1960:91). The plant grows in the upper valley.

Acalypha macrostachys Jacq., *pispita*, is a tall shrub growing at elevations from 2,800 m to 2,400 m. The branches are made into brooms and baskets, particularly the *isanca* used for straining chicha.,

Manihot esculenta Crantz., *yuca* or manioc, is an outstanding crop plant in the valley partially because it produces well with so little care. Some individual plants have as many as 20 edible roots and these roots do not have to be harvested at any particular time. If left in the ground 3 or more years, the roots grow very large, especially if the branches are pruned. A serious disadvantage of manioc is that the roots begin to harden upon being exposed to the air. Furthermore, manioc is virtually all starch with less protein than white potato or sweet potato.

Although all varieties of manioc are known to contain cyanogenetic glucoside, the amount in the so-called 'sweet' cultivars is sufficiently low that the juice need not be pressed out of the root. The Vilcanota Valley has only sweet cultivars, a pattern which this area shares with most of the Amazon drainage basin in Peru.[28] At least six important manioc cultivars are in the lower valley and perhaps more. The most important is *pasquito* with a pink epidermis said to preserve better outside the ground than the other cultivars. *Yanañawi* is much appreciated as a 'floury' kind. Other cultivars are *barilla* with a brown epidermis and *panti* with a reddish epidermis. *Motochi* (or *kello rumu*) is unusually large, and yellowish inside

[28] WEBERBAUER (MACBRIDE Vol.XIII,Part IIIA: 169-171) claimed that bitter manioc was grown from 1,500 to 2,200 m in Peru and that the poison had to be removed by special processing. This statement is erroneous for the Vilcanota Valley, as no bitter manioc is grown.

rather than white as are the others. Many chacras in the Chahaures area grow other cultivars (e.g. *taganero, yuca de arrocha* and *chillpi*) which may be more recent introductions from the lower Urubamba or even the Ucayali region.

The effective upper limit of manioc cultivation in the valley is 1,700 m and the absolute limit is at 1,980 m, but at one time or another it has been tried with minimal success up to 2,380 m. At its present absolute limit, this cultigen is definitely depauperate: 1 meter tall with small leaves and thin slender branches. Beginning at 1,500 m manioc is the most important root crop grown, and below 900 m it is especially significant as a food and as an element of the vegetative landscape.

Manioc is prepared for consumption in the valley in a limited number of ways. Unlike in Caribbean countries, no 'cazabe' bread is made, nor is Brazilian *farinha* (meal) prepared. Most commonly the root is simple boiled and eaten in a stew with other root crops and, rarely, meat. Manioc is also baked in ashes, a palatable, if bland, food. The preparation of starch is still known. Below Quillabamba, the country people prepare a fermented manioc beverage (*newayasiri*), a use of manioc which may have been learned from the forest Indians farther down the valley. The Machiguengas still masticate manioc to promote fermentation, but the Highland Quechuas in the Valley cook and grind up the manioc and add yeast from maize chicha to begin the fermentation.

Manioc is more than food. Because of its umbrella-shaped digitate leaves, the plant is used as a temporary shade for young coffee and coca plants. Thus manioc provides a return from the land before the tree crops begin to produce. The branches of manioc are also used as fuel and the leaves are sometimes fed to livestock.

Manioc is raised mostly for home consumption and all rural people in the lower valley grow it. Though sold daily in the Quillabamba market for the dwellers of the town, it rarely moves outside the lower valley because it cannot stand the cost of transport, does not keep well once out of the ground, and cannot compete with the plentiful starch foods grown at higher elevations. The pattern of use was probably similar in Inca times. The Incas certainly knew manioc (Quechua word: *rumu*), but with arracacha, potato, añu, oca and ullucu as highland sources of starch, probably manioc was a strictly local crop that only on occasion reached the city of Cuzco.

Ricinus communis L., *higuierilla* or castor bean, is of African origin and volunteers vigorously all over the American tropics. It is found where man has opened a new habitat, as on roadsides or around dwellings. Castor bean occurs in the valley below 2,700 m, but it is conspicuous only below 1,500 m. The peasants regard it as of no special value though they collect and burn the larger stalks, more than 8 centimeters across. Although the castor bean plant is not a particularly high quality firewood, it is readily accessible along roadsides and trails. The seeds are recognized as a strong purgative, but are cautiously used, for an excess dosage can cause death. Oil, pressed from the seeds was, until the early years of this century, used to lubricate the machinery in the sugar cane mills, a use that was known also on coastal plantations (TSCHUDI, 1847:129). There are no oil lamps in the valley;

candles still furnish light except in Quillabamba which now has electricity.

Jatropha Curcas L., *piñon*, a common shrub found widely in the American tropics, grows on disturbed sites in the lower valley. The stalks are used as fuel and the seeds are used as a purge.

Jatropha ciliata M., *wanarpo*, is the best known aphrodisiac in the folk medicine of the valley. The love potion is concocted from the juice of the stem and The Indians and cholos seem to be convinced of its efficacy. The plant itself does not grow in the valley now and must be obtained from lower elevations by trade. Wanarpo may have been one of the items traded from the forest Indians to the Incas, for its properties were known (PACHACUTI 1927) before the Conquest.

ANACARDIACEAE

Schinus molle L., *molle* or pepper tree, has been one of the most useful wild native trees in the culture history of the valley. In Inca times it apparently was a cultivated species along roads and near palaces; the Incas also used the tree's resin for embalming material (COBO:I:268). Parts of the tree have also had considerable medicinal value, for pain and swelling (CIEZA DE LEÓN, 1883:397), malaria (COBO: I:268) and many others maladies. After the Conquest, great numbers of large molle trees were cut down to make charcoal (GARCILASO DE LA VEGA 1960:274). Also, lye was made from molle ashes to promote the transformation of molasses to granulated sugar (PÉREZ VIDAL 1954:93).

Today in the Vilcanota Valley the molle tree occurs as a conspicuous element in the landscape from 3,300 m to 2,600 m. The species has a weedy tendency and many barren dry slopes and roadsides are covered with it. Very few of these trees are over 6 meters high, however, because they are cut down for firewood before they are very tall. In the domestic economy the ashes of the wood, rich in alkali, are used to remove grain hulls. Formerly the ashes were also used as a dye mordant. In some parts of the Sierra, e.g., the upper Apurímac Valley, *chicha de molle* is made from the pungent berries, but this drink, although known about, is no longer made in the Vilcanota Valley. Firewood for those Indians who cannot afford to purchase eucalyptus wood is the main use of molle today.

Mangifera indica L., *mango*, a domesticated fruit tree native to India, has become an important plant in the valley below 1,500 m. The tree has an attractive, symmetrical canopy, dense green foliage and grows as high as 15 meters. The pendant fruit is somewhat oval with a fairly thick yellowish-red skin when ripe, a large seed, and a fibrous but very sweet flesh. The mango does not yield fruit where the average monthly temperatures are less than 15° C. and therefore it is rare in the ceja de la montaña zone where the tree itself can survive. Below 1,500 m (Fig. 41), the mango yields prolifically and provides a cheap fruit which is consumed by all social classes. As in India, it thrives in the wet and dry seasonal precipitation regime and on the well-drained valley slopes.

The first mango seed is said to have been introduced into the valley in 1876 (GOHRING, 1934:31). This date seems fairly accurate as no travel descriptions written in or before the nineteenth century mention the mango. In less than 100 years, the mango has become well integrated into the folk agriculture in that section of the valley where it can grow. Most peasants have one or several trees near their huts. For reproduction, the seed itself is used rather than cuttings. In February the mango ripens and at most highland markets it is conspicuous and very cheap.

SAPINDACEAE

Sapindus saponaria L., *sulluku*, a wild tree of the lower valley bears red fruit containing saponin. At one time the fruits (and apparently also the seeds) were used as soap to wash clothes, but with the introduction of detergents, they are no longer used in this manner, as far as I was able to determine.

Dodonaea viscosa (L.) Jacq., *chamana*, is a small tree with very resinous leaves found between 2,700 and 2,400 m. The wood and leaves are valued as fuel. Formerly the ashes were used as a dye mordant.

RHAMNACEAE

Colletia spinosissima Gmelin., *rokke*, is a spiny shrub found growing in the upper valley from about 3,400 m to 2,600 m. It has a well-established use as fuel, particularly for bread ovens. The even heat obtained by burning this shrub is believed to improve the flavor and quality of the bread, and many small bakeries in the valley employ Indians to gather it from the hillsides. Because of the formidable spines, crude fences of rokke branches deter entry into chacras. The bark of this shrub is also said to serve as a substitute for laundry soap but commercial detergents now generally are used instead.

VITACEAE

Vitis vinifera L., *uva* or the Old World grape, has never been a succesful cultivated plant in the Vilcanota Valley, a fact which must have greatly disappointed the Spaniards who had visions of great vineyards in the Province of Urubamba section of the valley. But the vines which grew produced only small harvests of very acid grapes. Diseases now abound, and probably were also present in the colonial period, yet vine cultivation was more common in the past than now. Furthermore, the Indians do not readily or willingly expend the close horticultural attention necessary for successful viticulture. Currently only at Hacienda Tarabamba (near the town of Urubamba) is more than one third hectare devoted to grapes. There they are now grown mainly for table fruit, although as late as 1954, over 1,250 liters of wine were made on the hacienda and even some distilled spirits. Few highland people drink wine. This may be partly explained by the inavailability of grapes, by the inferior quality of the fruit, and by the competition offered by maize chicha which is much cheaper.

MALVACEAE

Gossypium barbadense var. *peruvianum* Cav., *utku* or *algodón del pais*, is, like all American cultivated cottons, a tetraploid, and it has a long prehistory of use in the Peruvian Coast (TOWLE, 1961:64-5). Cotton cloth was kept in warehouses in Cuzco in the Inca period, but whether the fiber was obtained from the lower Vilcanota Valley is difficult to know. We may assume that textiles of alpaca and llama fleece were more practical than cotton cloth for the mountain dwellers.

Whatever the nature of cotton growing and use in the past, today native cotton is grown in a casual manner around the huts at elevations below 1,500 m. Unlike cotton in North America, the plant is treated like a perennial. Care is taken to sow it in August so that it can be harvested the following August; if it matures during the rainy season the bolls turn yellow from the high humidity. A fairly common hereditary variation in the valley is the occurrence of brownish (*vicuña*) fibers caused by a mutant gene for lint color. Both the white and brown fibers are made into yarn or mattress stuffing. Most country people in the lower valley now wear machine-made textiles, so both spinning and the cotton plant itself have greatly decreased in importance since the 1930's.

In 1914 and again in 1918 Egyptian cotton (same species but different variety) was planted in the valley (BUES, 1917:47), particularly on Haciendas Echarati, Potrero and Huadquiña. This venture was short-lived, however, for low prices, coupled with low yields, discouraged further plantings. The distribution of rainfall in the valley is such that, unlike the native cotton grown, the Egyptian cotton had to be irrigated. Also, because the growing period of Egyptian was not in harmony with the rains, the fibers rotted in the bolls.

BOMBACEAE

Ceiba publiflora (St.Hil.) Schum., *ceiba*, is a tree of the lower valley with a thick trunk and protuberances on the bark. The 'silky' threads from the fruit pods are used as a packing material, and there is some vague suggestion that it has been used for thread but this was not possible to substantiate.

Ochroma lagopus Swartz, *chawar* or balsa, is a wild tree with a remarkably light and buoyant wood which has occasionally been found below 900 m in the valley. It is now considered rare, but the tiny name of the settlement Chahuares (meaning balsa trees) suggests that this may not always have been the case. The Incas may have used rafts of balsa wood in their attempts to penetrate the lower reaches of the Urubamba Valley. Strips of bark (COBO:I:283) were removed from the tree to wrap bundles of coca; today banana fibers are used for this purpose.

STERCULIACEAE

Theobroma cacao L., cacao, is a domesticated tree of considerable economic importance in the lower valley. In the first 400 years after the Conquest, cacao was almost exclusively an estate product. Only the haciendas were sufficiently well organized to put it on a commercial basis; there may also have been some Indian resistance to its cultivation. In the latter half othe nineteenth century (MARCOY, 1875:355-6), several haciendas became well known for their high quality cacao, and that (PAZ SOLDÁN, 1862:377) from Hacienda Echarati was touted as the 'best cacao in the world.' La Convención still remains one of the leading cacao producing areas in Peru.

Cacao is clearly a New World plant with two basic kinds (CHEESMAN, 1944) defined from tropical continental America: *criollo* and *forastero*. Within each group are many cultivars, including hybrids between the groups. Criollo cacao has been cultivated since pre-Columbian times in Mexico (where it was highly valued in Maya and Aztec civilizations), Central America and as far south as Colombia. Forastero cacao, on the other hand, originated in the Amazon Basin and, as far as can be determined, had no pre-Conquest associations as a cultivated plant. Later, however, in the eighteenth century in Brazil and the nineteenth century in Ecuador, forastero cacao was the stock from which large cacao plantings were made; it was also carried to West Africa and today is the basic type of cacao traded in world commerce. In Peru, the absence of cacao in the archeological record suggests that it was not known as a cultivated plant, but may have been a wild-growing curiosity. The Spanish chroniclers do not mention it in relation to Inca culture and there is no known Quechua name for it. It is likely that the Spaniards themselves brought cacao to the Peruvian Montaña in the sixteenth century. The earliest reference to it is by COBO (I:69) who in 1610 noted cacao as one of the products of the Cuzco yuncas.

Two kinds of cacao have a history of cultivation in the Vilcanota Valley: *chuncho* and *común*. Chuncho cacao has large, long rugose pods, reddish in color and aromatic, whereas común cacao has violet-colored, small and almost spherical pods. Chuncho is a better cacao and grows at higher elevations, but común is higher yielding and more resistant to disease. Although chuncho belongs to the forastero group, it is not clear if común is also forastero or a cultivar in the criollo group. Apparently the first reference to two kinds of cacao in La Convención is in RAIMONDI'S (1898:44) account of his travels in the mid-nineteenth century. The común cacao he noted may have been the same kind that was present in 1610 and may have been brought by the Spaniards from northern South America. Chuncho cacao was brought to the valley from the lower Urubamba drainage and was first cultivated at Hacienda Illipani, an estate just outside the study area below the Yanatile junction. The actual seed transfer to the highlanders may have been made by the peripatetic Piro Indians on their yearly trading expeditions more than 300 kilometers upstream, rather than from the nearby Machiguengas, although the latter tribe has a name for cacao (*sarwinimike*) which suggests that they were acquainted with it.

Común and chuncho or hybrids between them remain the important cultivars in the valley today, although they suffer from several diseases, mainly witches broom (*Marasmius perniciosus*), which reduce the yield. The upper limit of cacao in the Vilcanota Valley today is 1,410 m, but it is important only below 1,200 m. Between 1,410 and 1,200 m the pods are much smaller than below 1,000 m, but they are said to contain more fat. The center of cacao growing in the valley has remained around Echarati (approximately 960 meters).

Cacao seeds are planted about three meters apart – 500 trees per hectare – and always in the shade of larger trees. As with coffee, protection from the sun's direct rays is regarded as essential for cacao to grow and yield satisfactory. COBO (I:258) commented that cacao in Peru was not shaded from the sun as it was in Mexico suggesting that this practice was adopted later. A cacao orchard demands pruning, weeding and sometimes irrigation from June to September if it is to yield well. The young trees begin to bear 4 years after being planted but the first good yield is when the trees are 5 years old. There is only one crop a year for the pods take almost a year to mature. Harvest begins in December and ends in April. The usual yield per harvest from each mature tree is from $\frac{1}{2}$ to almost 2 kilograms of dry cacao. The pods are removed from the trees with a *palka*, a long pole with an iron hook at one end. The fruits are opened with a machete and the seeds, from which chocolate is made, are scooped from the pulp with a spoon-like instrument (*wek'ena*). Knowledge of fermenting the seeds appears to be lacking. The seeds are immediately sold to middlemen who take them to Cuzco to two small chocolate factories which supply all of southern Peru; some of the best cacao, however, goes to Lima.

Except among those who grow cacao, chocolate is not part of the peasant diet. The Indians and cholos regard cacao mainly as a cash crop for which they get money to buy necessities. A useful byproduct is the juice inside the cacao pod – known as *chicha de cacao* – which is collected and drunk. The emptied pods of cacao serve as fuel and the ashes are collected and moistened to form a chunk, a bite of which is chewed with coca when other ashes are lacking. The outer covering of the seed is thought to be a good cough medicine. Inside the pod and united to the seed are filaments which are dried and sold to the Indians of the Altiplano as a folk remedy.

THEACEAE

Thea sinensis L., *té* or tea, was introduced into the Vilcanota Valley in 1913 directly from Japan by Benjamín de la Torre, an hacendado in La Convención.

Commercial tea production in Peru began at Hacienda Maranura and Hacienda Huiro (the latter estate outside the valley proper) with the vegetative cuttings from these first plants which were of the *China bohi* cultivar. Today most of the tea is produced in the nearby valley of the Rio Lucumayo (or Amaybamba)(Fig. 3); in the main valley of the Vilcanota, only 60,000 kilograms of tea leaf are produced a year which is less than one-tenth of the total La Convención production (figures from Cuzco branch, Ministry of Agriculture). The bulk of the valley tea plantings are at Hacienda Mandor Chico (1,940 m) in the heart of the ceja de la montaña, where some 100,000 tea plants are picked. Formerly Huadquiña and San Pablo had large plantings but neglect and a shortage of pickers forced the owners to abandon them. The most favorable climatic zone for tea in the valley is in the moist section from 2,000 to 1,500 m, but the scarcity of even gently sloping land precludes much commercial planting beyond that which exists.

Tea production is now by the true seed which is started in seed beds, transplanted to nursery beds and eventually moved to previously cleared fields. As with coca cultivation, manioc is planted to protect the young tea plants from the hot sun. Irrigation water is applied during the 4 months when there is less rainfall. Although 18 harvests per year are theoretically possible, the yearly average is closer to 15. In the rainy season, November to March, the harvest takes place every 15 days; from April to October, every 25 days. Tea leaves are picked in the same way as coca, but unlike coca, tea must be dried in specially constructed sheds.

BIXACEAE

Bixa orellana L., *achiote* or annatto, a small tree native to tropical South America, is often considered to be a true domesticate. Inside the spiny brown pods is an orange-red fleshy covering on the tiny seeds that has been used as a folk dye all over tropical America. The forest Indians of the Amazon Basin have used it to paint their bodies for adornment and possibly to keep insects away. The Highland Indians clearly learned to use this plant from the forest Indians to the east (QUEVEDO, 1900:499), and most likely the Incas received achiote as a trade item until they learned to grow it themselves. The nearby Machiguenga Indians call the plant *potocri*, whereas the Inca name was *mantur*, a term which has now been replaced by achiote. Many places of certain Inca occupance below 2,200 m in the various eastern Andean valleys have the name 'Mandorpampa' (mandor = mantur); in the Vilcanota Valley, below the Machu Picchu ruins, is a place known by this name, but achiote trees are now absent.

According to COBO (I:254-5), achiote was used in food to give color to chocolate and stews, and in medicine to cure blood clots and promote urination. The Andean peasants today use it, as far as I was able to determine, only to color food. This suggests that the use of this plant as a coloring agent is very deeply rooted, and possibly in Inca times this red dye was a blood symbol which had religious significance. Today food is dyed most commonly for certain religious festivals.

After the Conquest, large quantities of achiote seed were exported from

America to Spain to be used as a dyestuff, and Peru supplied part of this demand. In the twentieth century, achiote has been used as a coloring for butter and margarine, and is exported via Lima to the United States, Netherlands, and other major dairying countries. The achiote production in the valley fluctuates from year to year depending on the price offered by the wholesale houses in Quillabamba. When the price is low, the country people often do not bother to pick it. However, because cancer may be caused by synthetic dyes, the demand for achiote has been high since about 1950. Nevertheless prices fluctuate greatly: in 1963, the price was S/200 per quintal (= 100 pounds), while in 1964, it was S/450 per quintal. Production rose 120 percent between these two years, not because of more plantings but by greater attention to its collection.

Achiote grows in the Vilcanota Valley below 1,480 m most commonly on the margins of coca fields or in the middle of such fields to cast shade. Although it can be propagated vegetatively, achiote is always grown from seed and not from cuttings. There is considerable variation in seed color, ranging from yellow to deep red ('*granate*') and the pod, while usually prickly, may also be smooth. A mature tree may produce up to 7 kilograms of seed per year. The harvest begins in August and continues until October. It is important that the seeds dry thoroughly on the tree before the pods are picked and taken to the patio. Women usually perform the tedious work of shelling the pods and winnowing the seeds. Before being taken to the wholesalers, the seed is still further dried.

PASSIFLORACEAE

Passiflora ligularis Juss., *tintin* or *granadilla* or passion fruit, is a domesticated climbing vine yielding a baseball-sized fruit with a tough but very thin covering. Inside are many small seeds embedded in a sweet, edible liquidy mass. The granadilla is probably native to the Central Andes although archeological evidence is rare. In the Vilcanota Valley, this fruit grows below 2,500 m and finds its most favorable environment in the rainy ceja. Although commonly in cultivation, it does volunteer. The fruit is an inexpensive item in many valley markets beginning in November. Here it is eaten fresh and not made into a drink as is done in Brazil.

Passiflora mollissima (H.B.K.) Bayley, *tumbo*, is a semi-cultivated twining species related to granadilla but with a different ecological niche. This plant produces a somewhat oval, bright yellow fruit with a more acid flavor than the previous species. Picked green and macerated, the tumbo is used by country people to coagulate milk for making cheese. It grows at elevations above 2,500 m and seems particularly common in Quispicanchis. The tumbo is much more casually grown, less highly valued, and less common in markets than is the granadilla.

CARICACEAE

Carica papaya L., papaya, is casually cultivated below 1,900 m in the valley around a hut or as part of a field border, but never in ordered large plantings. It readily volunteers and discarded seeds will often sprout and within 2 years bear fruit. The fruit is usually 20 centimeters long but occasionally may be as large as 40 centimeters. It is always eaten fresh, preferably sprinkled with lime juice to bring out the pungent flavor. A ripe papaya is so soft it cannot be handled without serious damage, and fruits destined for sale in peasant markets are thus picked before ripe. Papaya's cultural-historical position in the valley is not clear because archeological evidence for it in Peru is lacking. The cultigen, however, may be native to the eastern slopes of the Andes.

Carica candicans, chamburu, is the dwarf papaya that COBO (I:239) spoke of, and which grows at higher elevations than the regular papaya. The chamburu fruit is yellow, the size of a giant lemon, more bitter in flavor than the larger, more heat-loving kind, and is served cooked. It is especially grown in small backyard gardens in the towns of Urubamba, Yucay, and Huayllabamba, but seldom is seen in markets. The chamburu is not particularly well adapted to the low temperatures of these localities, for its leaves commonly show frost damage in June and July.

LOASACEAE

Loasa ranuncilifolia H.B.K., *orko kisa* a nettle-like herbaceous plant growing wild, is used externally to treat sharp pains.

BEGONIACEAE

Begonia fagoproides DC, *achankaray*, is an herbaceous wild plant common in the ceja, 2,300 m to 1,600 m. The ornamental flowers are admired and the rhizome of this plant serves as a famine food for the peasants.

CACTACEAE

Opuntia exaltata Berger, *pata kiska*, the most common cactus in the valley, ranges from 3,750 m to 2,900 m. This many-branched, red-flowering cactus normally attains heights of less than 3 meters and has formidable spines. It grows wild but it is also cultivated around human settlement. This cactus is frequently planted on the perimeter of chacras or on top of adobe walls to deter entry. Its usefulness is more than just as a living fence, however, for the spines are used as needles and the trunks of larger specimens are used as construction beams.

Opuntia floccosa Salm, *warako*, is a small earth-hugging cactus with yellow flowers, yellow-green fruit and covered by white hairs. It grows at high elevations, commonly above 3,700 m, and the small fruit is collected and eaten.

Opuntia perrita Griffths, *unkukyok kiska*, is a somewhat smaller version of pata kiska but with a denser covering of spines. It is found both cultivated as a living fence in the upper valley, and also wild-growing.

Opuntia ficus-indica Mill, *tuna* or Indian fig, native to Mexico, was brought to Peru probably after the Conquest. Unlike other cacti in the valley, it is never truly spontaneous and is always found near settlements. It bears an edible fruit known as tuna, sweet, juicy, and full of seeds. The original motive for its introduction may have been to serve as the host plant for cochineal, a scale insect from which a brilliant red dye (known in Peru as *macnu*) has been derived (GADE, 1972c). COBO (I:69), referring to the yuncas of Cuzco in 1610, lists 'cochinella' as one of its products. Although this is no longer the case, the tuna cactus has remained, valued now for its sweet fruits which are sold cheaply in markets from December to March.

Opuntia soehrensii Britton & Rose, *ayrampu,* a prostrate cactus with yellow flowers, grows wild throughout the upper valley and is also planted to form a fence around farmyards. It is still valued for its seeds which yield a reddish-purple food coloring which is said to formerly have been used also to dye textiles. Ayrampu seeds are sold in peasant markets.

Trichocereus cuzcoensis Britton & Rose, *hawak'ollay*, is a columnar cactus 5 to 6 meters high with few branches, greenish-white flowers and yellow fruit. It grows from approximately 3,600 m to 2,800 m. The peasants in this part of the valley mix the gum-like substance (*nopal*) produced by the plant with gypsum to make a whitewash. This probably started after the Conquest when Spaniards introduced the practice of whitewashing buildings. However, separately the gummy substance can also serve as a glue, for which the Incas may have used it. This large cactus bears edible fruits, called *peskemanka,* and is planted to form a living fence.

Cereus trigonodendron Schumann, *yunka hawak'ollay*, is another columnar cactus with smooth, light-green trunk, few spines and yellowish flowers and fruit. It grows from 1,600 m down valley, but appears to be in an unfavorable competition with the more mesophytic surrounding vegetation; perhaps it is a relict of a former dry period in the valley. The fruits are consumed fresh.

MYRTACEAE

Eucalyptus globulus Labill, *eucalypto* or blue gum, is today one of the 10 most important cultivated species of plants in the valley, for it is now the only significant source of wood above 2,500 m. Eucalyptus was introduced into Peru in 1865 from Europe; about 1880 (HERRERA, 1921:149), it was introduced into Cuzco and subsequently diffused throughout the temperature sections of the Department. By 1900, it probably was well established in the Vilcanota Valley above 2,600 m. Thus, within the short period of 75 years, this tree has become an integral part of the cultural landscape of the valley between 3,700 and 2,400 m above sea level. Below this elevational range, the trees are spindly because of high temperatures; above this range, the young trees are killed by low temperatures. Eucalyptus occasionally grows above 3,700 m but only within adobe walls which shelter the trees from the strong, cold winds. The absolute upper limit, above which no specimens are seen anywhere, is 3,875 m.

E. globulus accounts for almost all of the eucalyptus in the valley, with only an occasional grove of *E. camaldulensis*, a less hardy species of very little economic significance. Thus with over 99 percent of the plantings being of only one species, it would seem wise to taxonomically diversify the plantings, for a virulent disease specific to *E. globulus* could wipe out the entire tree population of the Sierra. A fungus disease that attacks this species is now known to be in the Department.

E. globulus was brought from the Iberian Peninsula where it became very popular after its introduction from Australia in the early nineteenth century (El Eucalipto en la Repoblación Forestal, 1956:262). It is native to south Tasmania and Australia south of latitude 37° where summer temperatures are moderate and the annual precipitation is above 1,300 millimeters. It survives temperatures as low as -6° C. In the Vilcanota Valley, the temperatures are roughly analogous to those in Tasmania, but the annual precipitation in the Sierra is much less. Thus in the dry season, May to October, eucalyptus seedlings and young trees need to be irrigated in the valley. Eucalyptus does not appear to volunteer although seed production is high; perhaps this is because the ground is very dry when the seeds are dispersed.

Individual peasants often start small plantations, and in Canchis, trees are protected by adobe enclosures. Since 1963 many communities have set out groves on communal grazing land with the assistance of the Governmental Forestry Service (Servicio Forestal). In the Indian community of Moccoraise in Quispicanchis, over 100,000 young trees will be used by all members on a share basis. The nursery beds are made by the Forestry Service; one is located in the valley near Yucay, Seedlings, $4\frac{1}{2}$ years old, are individually transplanted from the nursery beds to groves in a horticultural manner. When eucalyptus is two years old, the hard narrow leaves replace the fleshy wide leaves; after this transition a tree can usually survive the dry season without irrigation. When the tree is 15 years old, the trunk has a diameter of 50 centimeters and is the best size for cutting. Many trees are cut after only 10 years of growth because of the pressing need for wood. Replanting is not necessary for eucalyptus sprouts vigorously from suckers, often five on one trunk. This characteristic, plus its very rapid growth, have made eucalyptus an inestimably important floristic element in the valley landscape. For this reason it is sometimes planted in groves on flat, irrigable valley land, though farmers complain that eucalyptus groves shelter birds that eat part of the grain crop.

Eucalyptus is of utmost significance today in the valley as a source of fuel and building materials. The great scarcity of native trees and absence of other planted species has placed a premium on eucalyptus wood for a wide range of necessary items from beams to spindles to coffins. More and more agricultural tools are made of eucalyptus as the wood of suitable native trees becomes scarcer. The eucalyptus is logged during the dry season and it is a common sight at this season in the valley to see two men standing on a platform laboriously hand sawing the trunk into boards. With an axe,thin trunks are chopped into smaller pieces for firewood. Even dried leaves and branches are used for fuel in cooking ovens and in roof tile refractories. The leaves also have authentic medicinal value and are used as a decongestant; for this purpose, the foliage of young trees is preferred because the essential oils are emitted freely, whereas in adult specimens the oil is stored in glands and thus is difficult to extract.

Eugenia oreophila Diels, *unka*, is a small, now rare, tree of the lower part of the upper valley (between 3,000 and 2,600 m). The very white wood is easily worked and used in cabinet making and for carving ladles and other domestic tools. The Incas made wooden cups (*keros*) from this wood (VARGAS, 1962).

Psidium guajava L., *guayaba, sawintu* or guava, is a small semi-domesticated fruit tree that grows below 1,600 m in the valley. In many situations it is hard to determine if guava is actually cultivated because it volunteers so readily. Most trees may not have been planted on purpose, but once started, have been given some measure of protection by man who eats the fruit. The guava fruit is quite mealy in texture, not highly prized, and rarely seen in peasant markets.

ONAGRACEAE

Oenothera rosea Ait., *yawarchunka*, has a high repute in folk medicine of the valley. It grows wild on hillsides in the upper valley. The leaves are used as poultices for bruises, and they are also taken internally for such maladies as pneumonia and changes in the blood.

Fuchsia boliviana Carr., *chimpu chimpu*, is perhaps the most important of the several species in the genus *Fuchsia* in the valley. It grows in wet places below 2,900 m and in the ceja de la montaña is extremely common. The brilliant red blossom is valued as an ornamental and the plant is sometimes transplanted from its native habitat to farther up the valley. The Incas painted stylized fuchsia blossoms on wooden cups (*keros*). The fruits of this herbaceous wild plant are said to be edible, but their use as food was not verified.

UMBELLIFERAE

Petroselinum crispum Mansf., *perejil* or parsley, is native to the Old World and was brought to the valley in the colonial period. Its leaves are used as a seasoning *per se*, rather than to garnish foods. It is one of the fresh herbs sold in peasant markets and is grown throughout the valley.

Apium graveolens L., *apio* or celery, is occasionally grown in dooryard gardens but it is cultivated for its seeds rather than for its edible stalk. The seeds are ground and an infusion is made to treat pains.

Foeniculum vulgare Gaertn., *hinojo* or fennel, is an Old World perennial growing up to 1 meter high and found in all parts of the valley except the puna. It is sometimes cultivated and sometimes volunteers on trash heaps. Fennel is well integrated into the folk medicine of the region. The flowers and fresh leaves are used with other herbs in infusions to cure kidney inflammations, cramps, bronchial cough and nervous disorder. The custom of mixing herbs is believed to be a practice derived from folk medicine in the Old World; the Incas regarded each plant individually as being a specific cure for a specific illness. The fruits of fennel are said to be used as a purge.

Hydrocotyle alchemilloides Rich., *matek'lo*, is one of several species in the genus common in wet places in the Vilcanota Valley, especially as a ruderal along irrigation ditches. Matek'lo has had a considerable reputation in native medicine since Inca times (GARCILASO DE LA VEGA, 1960:92). The fresh leaves are applied as a poultice to the forehead to cure headaches and neuralgia as well as boils. Used in infusion, it functions supposedly as a diuretic and as a remedy for liver disease.

Daucus carota var. *sativa* DC, *zanahoria* or carrot, was introduced in the sixteenth century (COBO: I:415). Today it is of real importance in the upper valley and is one of the most commonly traded products in local markets. In this respect it is the only Old World root crop (besides onion and garlic) which succeeded in the upper valley. Perhaps its resemblance in flavor to arracacha made it more readily acceptable to the native peoples, but it would not have succeeded if both had the same ecological niche. The carrot grows as high as 3,700 m and therefore does not compete with the native New World umbelliferous crop. The roots are always cooked in stews and never eaten raw.

Arracacia xanthorrhiza Bancr., *virraca* or arracacha, an Andean domesticate grown in the rainy warm temperate section of the valley, is the most important root crop grown between 2,390 m and 1,800 m. It will successfully grow at higher elevations, up to Urubamba (2,880 m), but potatoes yield much better at these higher elevations and so little arracacha is planted. At elevations below 1,500 m the plant does not thrive because of the heat and fungus diseases but it is occasionally grown here. Few cultivated plants have such a well-defined ecological niche in the valley as arracacha has; it is the only root crop that is completely at home in the ceja de la montaña, with its rainy warm climate. Arracacha has not been commercialized and it is not an important market item. One rarely sees it anywhere in the valley except in the ceja zone.

The first mention of arracacha was made in 1533 by the soldier Diego de Molina who, upon returning to Spain, informed Oviedo in Santo Domingo: 'Hay muchas y buenas zanahorias de los morados' (HERRERA, 1941:4) ('There many good carrots grown by the inhabitants'). *Zanahoria* is the Spanish word for carrot and arracacha resembles in leaves and flavor the true carrot (Fig. 57). Unlike Colombia (where the plant may have originated as a crop: HODGE, 1954), southern Peru has relatively few cultivars and few uses for arracacha. Basically three cultivars grow in the valley. The best liked and the one considered to be the sweetest is yellow fleshed (*kello*) and often can be harvested 6 months after planting. The two other cultivars are *yurak* which has white flesh and matures in 9 months, and *kulli* which has a white skin but purple flesh. Although the roots of these last two keep better out of the ground, they are not considered as tasty.

Arracacha is propagated vegetatively by shoots (*hijos*) that grow from the main rootstocks. Because the shoots rot quickly in the very humid atmosphere, they must be planted shortly after harvest in October. They are placed in rows like potatoes. In the valley, arracacha flowers but does not set viable seed. In the harvest, the whole plant is dug up and the roots are cut from the main rootstock (*cepa*) which is inedible.

In Colombia the roots are used to make starch and to prepare a fermented drink while the leaf petioles are eaten as a salad (BUKASOV, 1930:128). None of these secondary uses is known in the Vilcanota Valley where only the fresh root is consumed. The peasants boil the root and add it to soups and stews, rarely do they eat it by itself. Apparently they do not roast or dry the root. Arracacha is a nutritious food with a high starch content and delicate flavor somewhat reminiscent of cooked carrots.

Cuminum cyminum L., *comino* or cumin, is a condiment that was introduced into the valley in the sixteenth century from Spain (COBO: I:417). Although the Spaniards used cumin a great deal, the indigenous population did not and today it is very rare in the valley, grown only sporadically in La Convención. Thus it is one of a large group of Old World plants rejected by the indigenous population and by subsequent generations of Spanish descendents. Nevertheless a few people continue to grow cumin as a curiosity.

Coriandrum sativum L., *culandro* or coriander, is a white-flowered herb of which all parts are used as a condiment except the root. This herb, introduced in the colonial period (COBO: I:419), is the seasoning most frequently used by valley peasants who put it in almost all soups or stews. Much coriander is grown below 3,000 m in dooryard gardens, and the fresh plant is frequently traded in local markets.

Fig. 57. Author standing in field of arracacha; in the background are Cavendish banana trees (1,940 m).

SAPOTACEAE

Lucuma obovata H.B.K., *lucma* or egg fruit (sometimes called canistel in English), is a semitropical fruit tree that grows up to an elevation of about 3,000 m in the Vilcanota Valley. It was domesticated in pre-Inca times and abundant archeological evidence of its presence is on the Peruvian Coast (TOWLE, 1961:77). The tree is about 12 meters high with shiny dark leaves. The fruit is ovoid with a dark green skin and dark orange flesh, mealy and almost flavorless. Reproduction is by planting the large shiny seed. Lucma may have been a common fruit tree in the valley in Inca times, but it could not compete with the fruit trees brought by the Spaniards. Today it is infrequently found in the valley, mostly as a yard tree behind adobe walls. Small quantities of the fruit appear only occasionally in the markets, and some informants claimed that they valued the wood of the tree – for cabinet making – more than they did the fruit. The fruit is also used much as a flavoring in ice cream.

LOGANIACEAE

Buddleia incana R. & P., *kishwar*, is a well-branched, medium-sized tree, 7-15 meters high. It has simple, opposite, dark green, leathery leaves which are whitish on the underside and red-orange flowers that grown in racimes. Kishwar grows from 3,750 m to 2,800 m in the valley (Fig. 43) and is especially predominant in the landscape above 3,400 m. Kishwar is highly valued for its wood in this wood-poor region. Now, as in COBO'S time (I:255), it supplies the basic material for agricultural tools. It is also much used for construction beams and is an excellent firewood, although less respected species are substituted for fuel if at all available.

Almost invariably kishwar is found near human dwellings and, at least in the Vilcanota Valley, is not found in the wild state. It is, in fact, a cultivated, if undomesticated plant, and it takes on special importance at high altitudes where sources of tool wood become critical. The tree is reproduced vegetatively rather than from seeds. After the trunk is cut, sprouts grow from the stump and there is no need to replant until after the second cutting. One can only conjecture how long this species has been in cultivation. A document dating from 1595 (ANONYMOUS, 1957:429) concerning woodlands of the Urubamba area indicated that kishwar trees were planted, and were not part of the natural vegetation (*madera brava*). It is very possible that kishwar was cultivated also in the Inca period.

Buddleia coriacea Remy., *kolle*, is a somewhat smaller tree than kishwar, but with longer leaves, denser branching and a knottier trunk than the above mentioned species. Like kishwar, its whitish-yellow wood is high quality and is used for essentially the same purposes. Kolle, however, can withstand stronger winds and heavier frosts than kishwar and grows therefore at higher elevations, from 4,300 m to 3,700 m.

GENTIANACEAE

Gentiana primuloides Gilg., *fallcha*, is a perennial, herbaceous wild plant that grows above 3,600 m in the valley. It is used in infusion as a sudorific, and its flowers are used in religious rites, as perhaps they were before the Conquest.

CONVOLVULACEAE

Cuscuta grandiflora H.B.K., *mallunwa*, is a parasite which attacks crops in the upper valley. The peasants regard it as one of the few truly noxious pests because it cannot be used as fodder and thus has no saving grace, but also because ingestion of any part of it is believed to cause sterilization of women.

Convolvulus crenatifolius R. & P., *willko*, is perhaps the most common of the convolvulaceous weeds that infest crop fields in the valley. These twining annuals are particularly successful in maize fields where they use the corn plants as supports. Although their presence is not appreciated, they can be used as fodder for cattle and other domestic animals.

Calonyction aculeatum (L.) House, *bejuco*, is the tropical moonflower vine that occurs in the lower valley. The dried vine of this plant is used in lashing construction beams together.

Ipomoea batatas Lam., *camote* or sweet potato, is of New World origin and was grown in the Andes before the time of Columbus. A subtropical plant, it does not thrive in temperatures much below 10° C., and its important distribution in the Vilcanota Valley is below 2,500 m. Although more nutritious than manioc, the sweet potato is, surprisingly, of little agricultural importance in the region. Although many peasants grow a few plants, hardly any depend on it for sustenance. At Santa Ana (near Quillabamba) O.F. COOK (1916b:87) remarked on the fact that the sweet potato was less important than manioc or uncucha, but far more important than achira. This statement remains true today.

Answers to inquiries indicate that diseases restrict the cultivation of this crop. However, another reason large quantities of sweet potatoes are not grown on any single chacra is that, like manioc, once sweet potatoes are harvested, they spoil quickly but, unlike manioc, they cannot be left in the ground until needed; they must be harvested when mature. In the Vilcanota Valley the tuberous roots are not dried, and perhaps they never were because the humidity is very high in that part of the depression where they are most grown. Sweet potatoes are boiled (*sancochado*) with other food, or occasionally, roasted over the embers.

The sweet potatoes in the valley may be classified as either sweet or starchy; curiously there is no inclusive term in Quechua for both kinds and the Spanish word *camote* (of Nahuatl origin) must be used to fill this terminological vacuum. The starchy kind is known as *kumara* and has an entire leaf, whereas the sweet kind is known as *apichu* and has a dentate leaf. Both kinds have several different cultivars, identified on the basis of root color, which may be white, purple or yellow. Kumara varieties are by far the more common of the two basic kinds in the valley.

POLEMONIACEAE

Cantua buxifolia Juss., *kantu*, is the best known indigenous ornamental plant in the upper valley. A shrub growing as high as 4 meters, it has brilliant red tubular flowers that are pollinated by hummingbirds. In Inca times, a stylized kantu motif was very common on wooden cups (YACOVLEFF & HERRERA, 1935:59-61). Today it is used in religious celebrations which may represent a holdover from pre-Conquest times. Kantu does not appear to be a wild plant in the valley, for it is invariably associated with human settlement; yet neither is it carefully cultivated.

VERBENACEAE

Lippia Fiebrigii Hayek., *cedroncillo*, is a native South American shrub that grows above 2,500 m in the valley. It is taken in infusion as a remedy for nonspecific ailments.

Lantana camara L., *tuknay*, is a common ruderal from 2,000 m down valley along roadsides and trailways. The leaves are said to be taken in infusion against stomach ache.

LABIATAE

Rosmarinus officinalis L., *romarino* or rosemary, brought from Spain in the sixteenth century (COBO I:405), has been adopted to some degree as a condiment by the peasants. It is sold in herb markets.

Mentha arvensis L., *hierba buena*, is a semi-spontaneous Old World mint found growing throughout the valley. It is commonly sold in markets to season soups.

Salvia tubiflora Smith, *ñukchu*, is a wild plant on the flanks of the valley at elevations between 3,700 and 2,500 m. The red tubular flowers of this small herbaceous annual have a very traditionalized role as an ornamental. In the Inca period the blossoms were copied on wooden cups, and today the flowers are collected from the hillsides to decorate the image of 'Our Lord of the Earthquakes', when it is carried around the city of Cuzco on Holy Monday (GADE, 1970b).

Minthostachys tomentosa (Benth.) Epl., *muña*, is the most important of several closely related species which are useful to the country people. This small herbaceous wild plant grows above 2,800 m. The highly aromatic leaves are stored with potatoes to protect them from insect damage, and they are also used as a seasoning as well as to make an infusion for nonspecific ailments.

Melissa officinalis L., *toronjil* or common balm, is an Old World plant introduced from Spain in the sixteenth century (COBO: I:413). It now grows spontaneously but may also be cultivated. Toronjil is used in folk medicine as a diuretic and in cooking as an herb.

Ocimum basilicum L., *albahaca* or sweet basil, is a common Old World condiment. It is traded in peasant markets in herb bundles and is seen most frequently in the lower valley.

SOLANACEAE

Nicotiana tabacum L., *tabaco* or tabacco, may have originated as a hybrid of two wild species in the lowlands near the Andean front (SAUER, 1952:48). The other cultivated species of tobacco, *N. rustica,* is also believed to be native to western South America. In pre-Columbian times, *N. rustica* had a far greater range in the New World than did *N. tabacum*, the tobacco of present world importance. After the Conquest, smoking became a widespread custom among the white population and the cultivation of the milder *N. tabacum* became profitable. Whether tobacco was smoked in the Sierra in pre-Conquest times is doubtful.

The cultivation of *N. tabacum* as a commercial crop in the valley below 1,500 m has been tied to governmental edicts. In the early colonial period, tobacco was grown here, but in 1745, the Estanco de Tobaco was created as a crown monopoly and by decree most tobacco was grown in the Zaña Valley on the north Peruvian Coast and at Chachapoyas, also in the north (CÉSPEDES DEL CASTILLO, 1954:145). After the 1824 revolution, tobacco again became a 'free' crop and most haciendas in the lower Vilcanota Valley cultivated it to sell. However, in 1910, the republican government reinstituted the monopoly and terminated tobacco growing in the valley. Some cholo peasants still grow it for homemade cigarettes and for medicinal purposes. Indians who chew coca do not usually smoke cigarettes and value tobacco as a poultice.

Nicotiana paniculata L. and *N. undulata* R. & P., both known as *k'amasayri* in Quechua, are the two most important ruderal species of tobacco in the upper valley. They are used as poultices and to cure aches and pains.

Nicotiana glauca Graham, *supay k'arko*, is a noxious ruderal that, according to RAIMONDI (1898:249) was accidentally brought to Cuzco by mule train in the mid-nineteenth century from the area around Cochabamba, Bolivia. The leaves have a deleterious effect if eaten by livestock. The peasants can ill afford to lose any domestic animals, and these plants are eliminated from the dwelling area whenever seen.

Capsicum spp., capsicum pepper, is the most important condiment of the country people, and widely grown and traded in the valley. Here, however, it is by no means used as generously nor in as wide a variety of dishes as in Mexico. However, fiesta foods are always highly seasoned with capsicum pepper. Of the three socio-economic groups, Indians use the pepper more than the cholos or mestizos.

HEISER (1953:225) believes that Peru has been at least a secondary center of speciation and domestication of this crop. In the Vilcanota Valley, two species stand out, *C. frutescens* L. and *C. pubescens* R. & P.; two other species known to exist in the Sierra, *C. annuum* and *C. sinense* may be present, but were not identified. *C. frutescens, uchu,* was probably brought to Peru from Mexico in pre-Columbian times (TOWLE, 1961:81). In the Vilcanota Valley it is grown in the dooryard at elevations below 2,800 m, but does best below 1,500 m. Reproduction is by the seed. Several cultivars are distinguished by color or size; among then are *pukauchu* which has medium-sized red fruits; *kellochu*, with medium-sized yellow fruits; and *chinchi uchu* with very pungent small green fruits. Fruits of this species are commonly dried (*chaki uchu*) and sold in this preserved form rather than fresh. Dried peppers are ground on a mortar to a powder (*allpa*) and added to many culinary concoctions.

C. pubescens, rokoto, is native to South America and is the capsicum pepper more commonly grown and used in the Vilcanota Valley. Unlike *C. frutescens,* the leaves and stems of this species are covered with hairs, but like the former, it has dark purplish flowers. The pungent fruit appears in many different shapes and several different colors, but the names of these cultivars were not obtained. Rokoto grows mainly in the upper valley, and the outstanding center of production is in Quispicanchis, especially near Urcos, from where it is marketed in fresh form throughout the upper valley.

Cyphomandra crassifolia (Ortega) Kuntze is a shrub or small tree which bears an edible, but somewhat astringent, egg-shaped fruit. Known as simply *tomate*, it is also called *yuncatomate, sacha tomate* or *berenjena* to distinguish it from the *Lycopersicum* tomate (the common tomato) which is known as *libratomate*. Tree tomato can grow below 3,000 m in the valley, but it yields best between 2,300 and 1,500 m. It is cultivated near dwellings or in gardens and is reproduced by cuttings rather than by seed. It does not seem to have dietary importance anyplace in the valley. It is sold in small quantities in folk markets. Cyphomandra tomato probably was domesticated in the Andes, for the genus has its greatest variation in this part of the New World tropics.

Datura arborea L., *kampanchu* or *floripondio*, is a beautiful solanaceous tree with striking white bell-shaped flowers. It is cultivated as an ornamental throughout the valley. In addition, the leaves are used as a poultice and the seeds may be used by some folks as a hallucinogen although no one admitted to doing it.

Physalis spp. includes two cultivated or semi-cultivated dooryard fruit-bearing plants which are said to occur in the valley below 2,500 m; both are named capulí, but have no relation to the capulí cherry except in so far as a Mesoamerican origin is likely for all three. *Physalis peruviana* L. is a perennial; *P. pubescens* L. is an annual. Both have edible but very small yellow fruits enclosed in a foliaceous husk. Neither species has much horticultural importance in the valley.

Solanum lycopersicum L., *tomate* or tomato, is very likely of Mexican origin (SAUER, 1950:521) and has almost no folk associations in Peru. Its pre-Conquest presence in the valley is doubtful, although VARGAS (1962:112) believes that Guaman Poma de Ayala's *ciccla yuyu* is the tomato. On the other hand, a subspecies of the tomato, the *pisko tomate (S. lypcopersicum* var. *cerasiforme*), is a conspicuous fruit in the lower valley. It appears to be both cultivated as well as a weed. The fruits are frequently sold in the Quillabamba market

Solanum muricatum Ait., *pepino, mataserrana* or *kachum*, is an herbaceous plant about 80 centimeters high which bears a fleshy fruit with a variegated purple and white skin. VARGAS (1962:111) identified a stone representation, found near Quillabamba in the valley, as that of pepino. In the Vilcanota Valley, from about 3,000 to 2,500 m, it is a casual dooryard plant and not grown on a commercial basis as it is in the Arequipa area. The pepinos seen in the Cuzco market come from Arequipa. The plant has lost the capacity to reproduce from seeds. The country people like pepino but say it does not grow well in the valley, and that it is susceptible to *soyra* (*Phytopthora infestans*), the same fungus that attacks the potato.

Solanum pseudolycioides Rusby, *t'ankar*, is a spiny shrub with hanging branches and bright orange fruit that grows as a ruderal in disturbed places between 2,400 m and the lower end of the valley. The seeds are said to be used as a purge.

Potato

Solanum spp., *papa* or potato, is the Andes' great food contribution to the world. Cultural (LA BARRE, 1947) and archeological (TOWLE 1961:86) evidence points to the Altiplano of Southern Peru and northen Bolivia as the probable place of domestication of the potato. In this region one finds the greatest amount of speciation and clonal elaboration as well as the highest degree of integration of this crop in the culture. However, a different hypothesis (SAUER, 1950:516-7) is that the potato originated as a crop before it reached the Altiplano, that it started as a diploid at lower elevations and, through genetic changes, became more resistant to low temperatures and thus could grow at higher elevations. Even assuming this to be a reasonable hypothesis, on the basis of cytological studies of potato clones in the Department of Cuzco no residual correlation can be made between chromosome numbers and altitude in the Vilcanota Valley (OCHOA, 1964). Ploidy level of a particular clone cannot be deduced from the morphology of the plant, and the sure method of determining chromosome numbers is to count them under a microscope.

The potato is grown mainly in high altitude places in the valley where grains do not thrive or where the potato is part of the crop rotation. In the valley more of the peasants' diet is derived from grains than from tubers, and grains bring higher prices than the potato. The potato nevertheless grows over a wide range, from 4,340 to 1,850 m; below 2,600 m it is peripheral. Above 3,800 m, the potato is sparingly cultivated but it is seen as high as the slopes above La Raya Pass (Fig. 58).

Potatoes are most important as a crop in the Canchis section of the valley, south of Combapata. Here they grow mainly on unirrigated land. Most of the technical agricultural aid in this province is directed towards improving the potato because agronomists believe it to be the crop best adapted to the environmental conditions which prevail there. A fundamental distinction is made in Canchis between 'sweet' and 'bitter' potatoes. Sweet varieties are usually consumed fresh although some are also used for chuño; on the other hand, all the 'bitter' varieties are made into chuño. On the valley floor both kinds are grown, but only bitter is

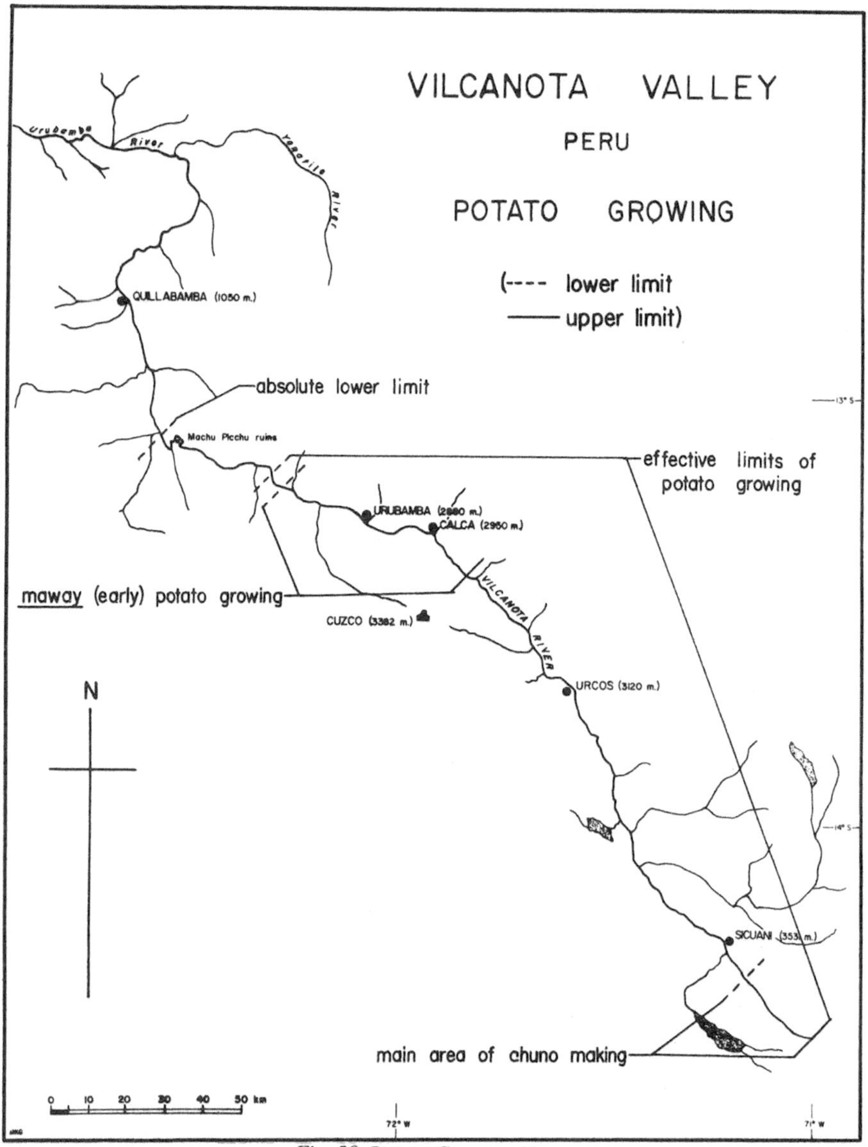

Fig. 58. Potato Growing Map.

grown on the upper slopes of southern Canchis.

In the Quispicanchis section of the valley, the potato is somewhat less important than maize as a crop. In Calca and Urubamba, the potato is grown in the valley mainly for the early harvest and thus, at this season, it is a highly commercial, market-oriented crop. The potatoes seen in markets in this part of the valley at other times of the year come from outside the valley. Trading valley maize for tubers from above the rim has characterized the economic geography of this part of the valley for a very long time. In part of the ceja potatoes are still cultivated (Fig. 34) on some chacras even though the yields are very small. Such attempts to cultivate a crop so ill-suited to the environment perhaps reflect the highland preferences of these peasants. The serranos of the valley below 2,400 m like potatoes very much, and if large quantities of the tuber did not move down the valley by railroad and truck, probably even more attempts would be made to grow them. In La Convención, the potato is not grown because high temperatures and humidity favor disease which make growing the tuber unfeasible. This is not to suggest that the potato has never grown here: DE CASTELNAU (1851:IV:286), travelling in the lower valley in the early nineteenth century, mentioned *pommes de terre* (potato) as one of the crops growing at Echarati (960 m). Formerly most potatoes for the lower valley were brought on muleback from the temperate areas to the west (Vilcabamba district); some still come from this area, but now the area within 40 kilometers of Cuzco supplies most of those that go to La Convención.

Potato Taxonomy of the Valley

An estimated 98 percent of the potatoes in the Vilcanota Valley, including those on the unirrigated upper slopes, are of the tetraploid series of *Solanum tuberosum*; the remaining 2 percent are triploids of the same species.[29] Within these two chromosome groupings ('ploidy levels'), there are many different potato cultivars. These cultivars are difficult to identify because most names are local and one cultivar may be known by different names even in neighboring villages. Also, in native potato taxonomy, cultivars are named solely on the basis of tuber characteristics, and the flower, leaf, or plant form are of no significance in the folk botany.

Varietal diversity of the potato in the Vilcanota Valley is greatest at altitudes above 3,450 m. Below this approximate elevation, no more than five cultivars are commonly grown, and one (*kompis*) is predominant, This suggests that where there is greater dependence on a particular crop, the variation within it is greater. In part such varietal diversity is a form of crop insurance, for if one cultivar fails to produce for climatic or pathological reasons, others are likely to succeed and a

[29] In the southern Peruvian Andes, according to DODDS (CORRELL, 1962), potato clones form a polyploid series from diploids to triploids to tetraploids to pentaploids, all within just three species. *Solanum tuberosum* includes three groups (2x, 3x, 4x), but most clones are tetraploids. *S. Juzepczukii* (3x) and *S. curtilobum* (5x) are highly resistant to frost, but otherwise not very desirable and so are cultivated only at extremely high elevations, mainly on the Altiplano.

crop is assured. Furthermore, each potato cultivar grown may have a slightly different use and/or flavor. Floury tubers are preferred over watery ones.

Many cultivars have a very local distribution and are never found much beyond the small area in which they apparently originated. Over many generations a taste preference for them develops so they continue to be grown, although their low yields and poor keeping qualities do not make them attractive to neighboring peoples. Certain cultivars have qualities which make them desirable to many Indians; only competition from similar kinds, or biological limitations on growth restrict the expansion of these potatoes into other areas.

In the Vilcanota Valley today cultivars of the kompis clonal group are clearly preferred. Kompis is the most common clone in the Department of Cuzco, and it appears to have been developed locally rather than introduced. In the valley it is grown from the head of the valley down well into the ceja de la montaña. Within the kompis clonal group are two outstanding cultivars, *kompis* and *alkay warmi*. Kompis cultivar is a small plant with many-branched stalks and light green, smooth-surfaced leaves. The flowers have a white corolla and a small, short calyx; the fruit is oval and contains abundant seeds. The tuber is round with deep eyes, smooth skin, yellowish flesh and purple sprouts. Kompis is very susceptible to insect and fungus disease. *Phytopthora infestans* commonly attacks this cultivar, as do larval worms. Indians say that potato worms are attracted to kompis because it is so 'sweet.' Compared with other cultivars, the yields of kompis are low. Its greatest virtue is its superb culinary quality; kompis is said to be starchier than most other varieties, better tasting, and is clearly the Vilcanota Valley peasants' favorite despite its deep eyes which make it hard to peel and cause waste. Closely related to kompis is alkay warmi, another popular cultivar especially in Urubamba and Calca. It has round tubers with deep-set eyes and red skin, white flesh, and pink sprouts. It yields about the same as kompis and is almost as good for cooking.

Less important cultivars are grown for specific purposes or because they fit into special ecological niches. *Kello wak'oto* has round tubers with deep-set eyes, yellow flesh, white skin, and pink sprouts. *Wayruro* and *paspasuncho* are frequently seen in the valley, particularly in Quispicanchis; the latter tuber is yellow, flattish, and with few eyes. *Bole* is a purple-flowered cultivar and is particularly abundant in Urubamba and Calca. It has a rather irregularly shaped tuber, sometimes oval, sometimes round, with red skin, deep-set eyes and a yellow flesh. *Kuchi aka* (literally 'pig dung') is so called because of its roundish shape and dark color reminiscent of pig scats. It is grown in the valley above 2,600 m. *Imilla* is grown in the valley above 2,800 m, but especially in the higher elevations of Canchis because it resists heavy frost. It is, however, not water resistant and thus is very susceptible to rot. Imilla is an Aymara name meaning girl, and may have been brought from Puno where it is the most common and best-liked cultivar in the Titicaca Basin. *Chakillo*, said to be very floury, is a notable cultivar in Canchis and Quispicanchis, and in one ayllu near Sicauni it is the only important variety. *Pokoya* is a potato consumed fresh and grown in southern Canchis; another important cultivar is *lomo* with a flattish tuber.

Potatoes grown specifically to be dehydrated have the general name of *papas amargas* ('bitter potatoes') and are grown on the cold upper slopes of the valley, mainly in Canchis. *Mayllku* is one of the most important of these. It is a rather large, flattish cultivar used for chuño and is found growing in southern Canchis, except in those places where the iron content of the soil is high. *Yana mayllku*, small and round, with a dark skin, is another bitter variety for chuño.

About 1950 the hybrid potato, a creation of modern science, was introduced into the Sierra and has been slowly gaining popularity. Estate owners quite commonly grow hybrid potatoes now, but the indigenous farmers and consumers generally have not accepted them. The most common hybrid cultivar in the valley is *mantaro*, a purple-flowered variety derived in 1949 from a cross between the varieties *jiruco* (from the Central Sierra) and *casablanca*. The tubers of mantaro are generally large and somewhat irregular in form, some being round and flat, others long and flat. The skin is whitish, the flesh is white and the sprouts are white with purple at the base. Its eyes are fairly deep. Mantaro is a semi-precocious cultivar of high yield, fair protein content, but it is somewhat watery. It adapts to a wide variety of climates and soils and stores fairly well. It is not very resistant to *Phytopthora infestans*. Its chief advantage is its high yield: mantaro yields 20,000 kg/ha, while kompis produces 15,000 kg/ha. To the peasants its chief disadvantage is its insipid flavor; they do not like it for a food. However, mantaro is said to make good chuño. Casablanca, one of the parent cultivars of mantaro, is also hybrid and some of it is grown in the valley, again, mainly on estates. The tubers of casablanca grow very large and although the Indians are impressed with their size, they do not like them because they are watery. Watery potatoes do not keep as well as the kinds that are high in protein.

Comprising several different cultivars is a special group known as chaucha potatoes which are believed to be triploids. All mature quickly and can be harvested within 90 to 105 days. Commonly they are grown for the early (*maway*) harvest in the Provinces of Urubamba, Calca and sparingly in Quispicanchis. They are delicate and do not resist either high or low temperatures; this, in part, explains their limited distribution. Relatively high in water content (compared with kompis), they cannot be stored a long time and must be consumed as soon as possible after harvest. Chaucha potatoes are always consumed fresh, usually boiled; they are never made into chuño. Two of the most important cultivars of chaucha potatoes in the valley are *pitikina* and *birundus*.

In the few places in the valley (or above it) where natural conditions are so rigorous only extremely frost-resistant cultivars can be depended upon. The cultivar *ruki*, believed to belong to the pentaploid *S. curtilobum*, is found in several tiny Indian chacras above the elevation of 4,080 m in the valley.

Potato Planting

Within the valley three potato plantings are made at different times of the year. The most important is the big crop (*hatun tarpuy*). Much less important is the early crop, or *miska maway*, and least important is the medium-early crop, the *chaupi maway.*

For the 'big crop,' the potatoes may be grown on either irrigated or unirrigated land. Irrigated fields are fertilized and the crops grown on them are rotated annually, commonly in a potato-legume-grain cycle. On unirrigated (*temporal*) land, potatoes are grown only 1 year out of 4 to 7; the land is left in fallow during the other years. In Indian communities some unirrigated slopes (*laymi*) are used for periodic potato cultivation and are still occasionally farmed communally. All members of the community participate in plowing, sowing, cultivating, and harvesting. One old man is selected to guard the fields against theft and to invoke the spirits and saints in case of frost, hail, or drought. Today most potatoes are grown on individually-owned and individually-worked land.

Cultivation

Plowing for the big crop is done as early as March, near the end of the rainy season to take advantage of the humid ground. Many Indian farmers believe the foot plow (*chakitaklla*) to be the only appropriate tool for preparing potato land, and often use it even when a plow and oxen would be more efficient. Most cholo farmers use the plow and oxen to plow potato land if the field is relatively flat and has rectangular dimensions. On the steep slopes of the valley, or on flat land that contains many large boulders, the chakitaklla is used by all farmers.

The big crop is usually sown in September, although above 3,650 m this occurs in October or November after the danger of frost. Many Indians carefully avoid sowing during a quarter or full moon because they believe that potatoes sown then will bring a small harvest. Before the sowing, the peasants make invocations to the Virgin of Nativity, Mother Earth and the tutelary spirits of the region. Libations and burnt offerings of llama grease, coca leaves, and incense usually accompany this ritual and it is customary for a potato farmer to put some potato propagules in a blanket, add a handful of earth, kiss it, and ask for a blessing.

The holes, about 30 to 40 centimeters deep, are opened with a lampa or raukana. A woman with a blanketful of potato propagules around her waist, drops them in these holes. The number of seed tubers for each hole depends on the size of the tuber: two, if large; three, if medium-sized; and five or six if very small. The practice of using very small tubers for reproductive stock partly explains the small size of the potatoes harvested. Sowing very small tubers discourages would-be thieves, but it also reflects the need to keep the larger ones for food.

Potatoes on flat bottom land are lightly irrigated when the tubers sprout, about 30 days after sowing. When the potato plant is 20-25 centimeters high, the fields are weeded for the first time and soil is banked around the roots; also those farmers who can afford it, apply manure or other fertilizer. The second earth banking again involves both banking and weeding, and is more carefully done, because by this time, the tubers have formed on the roots. All potatoes grown under irrigation are given two aporques; on unirrigated slopes, however, there is but one, usually around the time of flowering. Anxiety about the success of the crop rapidly increases around blooming time. Homemade crosses, called *papa kawaktaytacha* ('Lord-who-protects-potatoes'), are placed in the fields against any unforeseen meteorological occurrences which might destroy the crop. Indians cut off the flower of the potato plant in the belief that this will increase tuber yields. Another folk belief is that if menstruating women walk through a field of flowering potatoes, the leaves will blacken and dry up and the tubers will be very small.

The potato plant has many enemies which often reach alarming numbers, and which destroy at least one fourth of the annual harvest and damage still another one-fourth. The major insect pest in the valley is the *papakuru* or *laka* (*Premnotryes latithorax*), a coleopteran larva. This fattish white worm with an orange head is found in the valley above 3,000 m in the Provinces of Quispicanchis and Canchis. It tunnels into the tuber and often half a crop of potatoes may be infested with the worm. Nevertheless, potatoes so damaged are eaten, not destroyed, and at least a few Indians even toast and eat the worms themselves. If the larvae become too numerous in the fields, many farmers stop planting potatoes for a period of several years (some say 7 is best), after which time the fields are said to be again free of the pest. This is not a superstition for some agronomists regard it as a very practical and efficacious control measure.

Another major potato pest in the *sillwi*, a lepidopteran caterpillar of which there are several species. This larva, about 5 centimeters long, fat, brownish-gray, attacks the root neck of recently sprouted potatoes causing the plant to wilt and die. Another larva is the *ichukuru* which mines into the leaf parenchyma, the stalk, the root, and superficially the tubers. The *illa* is a long dark-green larval worm which eats potato leaves. At least three beetles damage the potato in the valley: the *chilliko*, a black one that eats the young leaves; the *k'arwa* that eats leaves, and the *rak'a* that eats the roots and tubers. The main fungus disease is *Phytophthora infestans*, a worldwide plague which causes the leaves to wilt and, as a result, the tuber yield is very small. Known as *soyra* or *rancha*, this fungus becomes more common as one approaches the humid atmosphere of the ceja and it effectively precludes potato cultivation on any serious basis at elevations below 2,600 m.

Potato Harvest

About five months after sowing, the potatoes are mature and ready to be harvested. In harvesting, one person may dig up all the potatoes of any one field, letting the other people collect them; or, several persons, each with a lampa or raukana, may dig the tubers, throwing them onto a common pile. Sometimes a pick is used for the digging. Whichever tool is used, as the tubers are harvested, the land is levelled. Peasant farmers get very low potato yields, usually only 3-4,000 kg/ha. On the other hand, the hacendado, using chemical fertilizers and bird manure from the coast, harvests 20-30,000 kg/ha. All of the potato plant is used by the peasants, except the fruits which are considered to be poisonous. The potato foliage and stalks are fed to livestock, even though it is not a preferred fodder because of its high percentage of potash. After the harvest, children comb the fields for remaining stalks which are gathered and used for cooking fuel.

A highly traditional part of the potato harvest is the on-the-spot roasting of some of the tubers right in the field. A *watia* (oven) is constructed of clods of dry earth, available only during the dry period. The clods are piled about 50 centimeters high, igloo fashion, and straw and wood from nearby hills are used for the fire which is built within the earthen oven. The proper heat for roasting the tubers is indicated when the clods become white hot. The ashes are removed with a stick and about a dozen newly harvested tubers are placed in the oven. After roasting for about 25-30 minutes, they are removed from the makeshift oven, allowed to cool somewhat, and eaten right in the field as the first sampling of the fruit of their labors.

After being harvested, the crop is carried from the field to an enclosure where the tubers are sorted according to size and variety If the potatoes have been grown as a community effort, the crop is divided among the members of the ayllu in the field. The largest tubers will be eaten fresh; the average-sized ones will be kept for seed if they have big eyes; and the very small potatoes will be made into chuño. Sorting may take several days; if so, each pile of tubers is covered at night with straw to insulate it from frost which changes the nature of the starch grains. The Indians poetically refer to this covering as 'not seeing the stars' (*mana koyllorpa rikunanpak*). The potatoes are generally stored in small adobe storage enclosures (*markas*) in the huts. However, at high altitudes, below ground storage is sometimes used: in a large pit, the tubers may be preserved for as many as 8 or 10 months. In storage, the potatoes are never allowed to touch the ground, for if they do, they soon begin to sprout. Sometimes they are placed on just a bed of straw or maize leaves, but usually they are stored on a layer of *muña* (*Minthostachys* spp.), a labiateous plant which seems to discourage fungi, bacteria and larva. I have also heard of green apples being stored with the potatoes (in addition to a layer of muña) to prevent wormy tubers; apparently as they ripen the apples emit a chemical that may impede larval development.

Dehydration

After the 'big' harvest (*hatun tarpuy*), a certain part of the crop is set aside to be made into chuño. In the Vilcanota Valley potatoes are dehydrated as low as 3,400 m, although only above 3,700 m is it common. It is done during the low sun period, June to August, when nights are coldest. Any potato can be dehydrated but some cultivars are preferred. All 'bitter potatoes' are so processed as are many cultivars of the 'sweet' group, such as kompis and the hybrids casablanca and mantaro. Actually two different kinds of dehydrated potato, collectively called simply chuño, are made and both by the 'freezing and squeezing' method: one is chuño or chuño negro, the other is tunta or moraya.

To make chuño or chuño negro, the tubers are spread out on the ground·fully exposed to the cold night temperatures of the Andean low sun period. Sometimes this involves moving the tubers to an especially cold spot in the hills where a small

Fig. 59. Chuno and moraya in an enclosure on the puna.

hut is erected to shelter the people doing this work. At temperatures of -4° C. the plant cell walls burst and the tubers become blackish and shriveled. In the morning, the Indians gather them into small piles and tramp on them with their bare feet to squeeze the moisture from the broken cells and to remove the skin. This procedure may be repeated each morning for 2 weeks until, with the aid of drying winds and a brilliant sun, the tubers become completely hard, dry, and blackish gray. The end product has only about one-third the size and weight of the fresh tuber.

The other kind of dehydrated potato is moraya which has a white outer covering and which requires greater skill and effort, as well as lower temperatures, than the more simply processed chuño. In the Vilcanota Valley, moraya is made only above 3,600 m, because only at these altitudes do the temperatures consistently fall to -8° or -10° C. At Occobamba, potatoes are subjected to a single night's frost and in the daytime, they are covered with straw to prevent them from being blackened by the sun's rays. They are then placed in the running water of the Vilcanota River for about 3 weeks. Finally taken out of the water, they are spread out and for about a week they are exposed to the night frosts, being carefully covered during the days. The last day, the Indians tread the tubers with their bare feet to remove the moisture content and then place them in the sun to dry for as long as 2 weeks. The end product is a tiny shriveled ball having approximately one-sixth the original volume of the potato (Fig. 59). Since the size of the tuber is so greatly reduced in moraya, it is somewhat of a luxury item, for in markets it costs twice as much as regular chuño. Moraya making is not common in the valley; the country people do not grow the kind of potato best used for moraya, nor does the temperature consistently reach the low levels necessary to make good moraya.

Two much less important ways to process potatoes result in *kachi chuño* and *kokopa*. The former is made only in May at the beginning of harvest when the frosts are still light. The starch grains are only partially changed, and the tuber cannot be preserved for more than several weeks. Kokopa requires no frost, but is prepared by boiling, peeling, and then sun-drying the tuber.

Maway Harvest

The sowing of early potatoes (*miska maway*) involves different practices and is done on a smaller scale than the main crop. Sowing occurs between June 15 and the end of July, and is often done on the feast of Saint Andrew (San Andrés) who is regarded as the patron of maway potatoes. The variety used for maway potatoes include several cultivars in the chaucha group which mature in 3 months. Those sown in June are harvested in late September or October; those sown in July, in late October or November. Potato supplies are low around the end of the year and the maway harvest of new potatoes (*mosok papa*) brings a high price. The large tubers are always sold while the farmers who grow them keep the small ones to eat, but within 60 to 70 days they begin to spoil. Maway potatoes are grown at elevations between 3,000 m and 2,700 m (Fig. 58). At higher elevations the low temperatures of the dry season freeze the plants unless they are in sheltered locales; and at lower elevations, below 2,700 m, weevils and particularly fungi, favored by the high humidity, reduce the yields below the acceptable level of profit. However, around the town of Urubamba, the high prices brought by the offseason harvest of the maway potatoes permit them to compete with maize for the fertile, irrigated bottom lands, where they are usually fertilized with barnyard manure.

Medium-early potatoes are sown in August and September and harvested in March; this is known as the chaupi maway harvest. Instead of chaucha cultivars, semi-late varieties are sown, for example, kompis, alkay warmi as well as bole, imilla and mantaro. This harvest is of small proportions compared to the maway harvest.

The potato as Food

The potato, fresh or dehydrated, is prepared for consumption in a surprisingly limited number of ways. Most commonly, it is merely boiled (*papawaiko*), and used alone or in a variety of soups and stews. Chuño and moraya are usually cut up in stews or soup. *Moraya phasi* is made by soaking it overnight in salt water, and filling the swollen moraya with cheese. Baked potatoes are not a usual manner of preparing this tuber except during the harvest when small earth ovens are built right in the potato fields. Fried potatoes, a post-Conquest method of preparation, are popular among mestizos and cholos, but not so much among Indians. Only infrequently are mashed potatoes prepared. Chuño and moraya are sometimes ground to make a mush eaten with milk or water. Potato starch is made, particularly in Canchis.

Saracha contorta R. & P., *awaymanto*, a solanaceous plant of the upper valley, has small edible fruits which are collected in February and March. Occasionally they are sold in markets.

SCROPHULARIACEAE

Calceolaria cuneiformis R. & P., *puru puru*, is a wild herbaceous plant found on dry hillsides in the upper valley. It is valued in folk medicine for its diuretic properties.

Escobedia scabrifolia R. & P. has been collected from the wild in the lower valley perhaps since Inca times. RAIMONDI (1898:51) noted that it was collected in the lower valley around the middle of the nineteenth century. The dried root of this native shrub, known as *palillo de raíz* (as opposed to *palillo chuncho* which is turmeric), is considered to be the superior source of yellow food coloring. In most highland towns in the Department of Cuzco small loaves of bread, dyed bright yellow, are traditional delicacies at festival times. *Escobedia* is rare today, perhaps because of intensive collecting.

BIGNONIACEAE

Tecoma sambucifolia H.B.K., *waranway*, is a small tree of the upper valley that is planted between 3,000 and 2,500 m in hedgerows. The wood of this tree is valued for agricultural tools, particularly scratch plows; large trees suitable for this purpose are, however, now rare.

Crescentia cujete L., *pamuco*, is a tree which yields large gourd-like fruits that are hollowed out to be used as vessels. It grows below 1,500 m in the valley, usually, but not always, in association with man. *Crescentia* fruits have been found in archeological sites on the Peruvian Coast (TOWLE, 1961:88), and it is likely that the tree grew in the lower valley in the Inca period.

Cybistax sp., *achiwa achiwa*, a tropical shrub growing below 1,300 m, is used as a diuretic by many peasants in that area of the valley.

Stenolobium stans H.B.K. (native name unknown) is a yellow-flowering shrub that is planted as a hedgerow component between 2,900 and 2,500 m.

PLANTAGINACEAE

Plantago hirtella H.B.K., *llanten,* a native herb found in moist sites, is important in folk medicine throughout the valley. Its leaves are used to treat a variety of maladies; among them bladder trouble, for which an infusion is taken.

Plantago major L., also *llanten*, is a somewhat larger plant than *P. hirtella* but with similar morphological characteristics and a similar ecological position as a ruderal in the upper valley. *P. major*, however, is an Old World plant introduced some time during the colonial period, and both the name and uses of this plant have been transferred from the native plantago.

RUBIACEAE

Genipa americana L., *witok* or genipap, is a tall tree valued not only for its strong wood used to make utensils, but also for its fruit. The pulp of the fruit is edible and the juice yields a deep bluish-black dye. Genipap occurs down valley from 1,500 m, and while I believe it to be part of the natural vegetation, it appears to have a protected status.

Coffea arabica L., *café* of coffee, was taken to Peru from Colombia in the eighteenth century and in 1784 was brought to the Vilcanota Valley from Huánuco in Central Peru (MENÉNDEZ RUA, 1948:33). As in most of the rest of the New World tropics, *C. arabica*, native to the Ethiopian Highlands, is the only species to have been introduced here. Coffee is grown in the valley between 2,100 m and 730 m, but that of high quality grows at elevations from 1,800 m to 1,000 m (Fig. 42).

Only in the twentieth century has coffee been seriously cultivated in the Vilcanota Valley. Methods have changed little in the past 150 years. Coffee is still propagated exclusively from the seed. This results in greater genetic diversity than would result from planting slips. Three or 4 seeds are placed in a hole and the weaker seedlings are later eliminated. Seeds are also started in nursery beds and then transplanted. On land which is newly cleared for coffee, some large trees are left standing to provide shade. They form a semi-forest habitat much like that in which the indigenous peoples of Africa grow their coffee. Coffee is also planted in coca fields when the peasants wish to replace coca with coffee; the coca is kept until the coffee trees start producing. This occurs after 5 years. Shade trees of the genus *Inga* are transplanted from nursery beds to these fields, for the Vilcanota peasants cannot conceive of growing coffee without shade. They firmly believe that beans shrink and become wrinkled and small if the coffee trees are exposed to the sun's full rays.

Other aspects of coffee cultivation include weeding the young seedlings and, in a few places, irrigation, although most coffee in the lower Vilcanota Valley is not irrigated. Coffee is susceptible to the ravages of the cutting ant which attacks the leaves in large numbers, and fungus root disease (*ayawaiko*) attacks the young stalk. Volunteer coffee trees are often seen growing sporadically on very steep hillsides; a bird which eats the coffee cherries, the *manak'arako*, may be responsible for this seed dispersal.

In the Vilcanota Valley coffee cherries are picked individually rather than by branch stripping. Coffee which is picked green is called *café de pozo* because it is put in a pit and trampled to remove the less pliable pulp. With ripe cherries, however, the freshly picked fruit is soaked in water 2 days to facilitate depulping. After the pulp is removed by a crude press device or a Bentall handmill made in England, the beans are immediately washed several times in running water. This removes the ivory-colored mucilaginous integument. Sometimes the depulped seeds are left in a tank of water for a day before the membrane is removed, but if left longer, the beans turn yellow and decrease in value. The beans are spread in the sun to dry for 6 to 9 hours and are later dried in the shade an additional 5 to 7 days.

Both estate owners and peasants grow coffee and today it is the most important cash crop in the lower valley below 1,800 m. Because coffee prices steadily increased in the 1950's and early 1960's, much land has been taken out of coca and other crops and devoted to coffee. Nevertheless, yearly fluctuations in price always occur so that too much emphasis on coffee cultivation can put growers in a very unfortunate economic position. Formerly when transport to market was by mules, much coffee was simply abandoned in the valley because the cost of bringing it to market was more than the going price. Coffee yields in the valley are very low compared with Brazilian yields; they average 275-375 kilograms of beans (*grano*) per hectare. A yield above 600 kilograms per hectare is considered to be very good. In large measure, the low yields result from no fertilization and the use of genetically inferior coffee stock, basically the same as that growing here in the late eighteenth century.

All peasants in the lower valley drink coffee daily. But it is by no means universally used in the upper valley; many Indians and cholos either cannot afford to drink coffee or they prefer herb infusions.

Cinchona pubescens Vahl., *cascarilla de Santa Ana* or cinchona, has been the most valuable of several wild species found in the valley that contain quinine in the bark, an effective anti-malarial agent. If cinchona was used as a febrifuge in Inca times, it was not for malaria which is probably post-Columbian in the New World. Beginning around the middle of the nineteenth century, much of the eastern Andean slope from Colombia to Bolivia was searched for this valuable plant. Although the highest quality bark came from plants on the Peruvian-Bolivian border area, the lower Vilcanota Valley below 2,300 m was also explored and exploited. CASTELNAU (1851:IV:53), in this travels in the lower valley in the 1840's, mentioned quinine bark being collected around Chahuares where it was stored in specially constructed huts. The bark was collected in the valley until the 1890's when the shrub became depleted. During World War II, the price reached very high levels and the bark was again collected for several years. Knowledgeable country people in the lower valley say that the current price of bark does not warrant the difficult task of collecting it from the almost inaccessible hillsides where it still can be found.

CAPRIFOLIACEAE

Sambucus peruviana H.B.K., *sauco*, is a smallish tree with pinnate leaves often seen cultivated as a hedgerow component, particularly below 2,800 m. The fruits of this tree are reported to cure sore throats.

CUCURBITACEAE

Cyclanthera pedata (L.) Schrad., *achokcha*, is an annual vine with digitate leaves that qualifies as a South American domesticate. It produces a peculiar hollow fruit with flattish black seeds. In coastal Peru, a cultivar known as *caigua* grows with a fruit 6-8 centimeters long and with a smooth surface. In the valley two cultivars are apparent: one has a smooth fruit $2\frac{1}{2}$ to 5 centimeters long; the fruit of the other is the same size but is also covered with soft protuberances that look like spines. These highland cultivars are fairly common in the valley and are casually cultivated in the same field with maize (Fig. 34). Achokcha has a wide ecological amplitude and is found in the entire valley below Sicuani (3,550 m). In no place, however, does it assume much importance as a crop and is, at best, regarded as a distinctly minor component of the agriculture. Occasionally Indian women will sell achokcha in the native market at Pisac to other Indians. The fruit is used as a boiled vegetable and in a stew with capsicum pepper.

Lagenaria siceraria (Moll.) Standl., *poro* or gourd, is an Old World plant which was cultivated on the Peruvian Coast before the use of pottery (TOWLE, 1961:93). On this basis, it is reasonable to speculate that it occurred in the Vilcanota Valley in Inca times and probably before. The gourd is an annual and the only cultivated cucurbit in the valley with a white flower. It was not seen growing above 2,300 m. The rind of the fruit is hard and makes superb containers and vessels for household use. Many cholos now buy metal and ceramic containers which replace gourd vessels, particularly in the upper valley where this plant cannot be successfully grown. Gourds were not seen in any markets, although formerly they must have been a significant item of trade.

Cucurbita maxima Duch., *sapallu*, has many cultivars differentiated on the basis of shape, size, and color. This domesticated cucurbit has kidney-shaped leaves, white or brownish seeds, and a soft peduncle. The fruits attain enormous size, are usually round, and are sold in pieces rather than whole. Sapallu is always eaten cooked, either sliced and boiled or mashed in a stew. This plant is a South American cultigen and on that basis probably has the longest history of any cucurbit in the valley. Today it is the most important cucurbit cultivated in the valley where it grows particularly below 2,600 m and rarely above 3,000 m.

Cucurbita moschata Duch., *lakawati* or *calabaza*, is a domesticated cucurbit of apparent Mesoamerican origin (TOWLE,1961:91). It grows in the Vilcanota Valley mainly below 2,500 m. Although there are many cultivars (or at least shapes and sizes), the common kind has whitish-green oblong fruit and whitish blotches on the leaves. The peasants do not particularly like this species and if they do eat it, it is used before it ripens. Once ripe, the fruits are fed to hogs. Nowhere in the valley was *C. moschata* seen to be seriously cultivated, and its casual use may in part be from volunteers.

Sicyos Baderoa Kook., *putak llanko*, is a wild native plant which grows in the upper valley. The seeds are used as a purge, but more important are the ashes from the leaves and stalks which are rich in potash and used as fertilizer. Those peasants in the Urubamba area who grow fruit in backyard enclosures fertilize their trees with these ashes.

CAMPANULACEAE

Lobelia nana H.B.K., *pilli*, is an aquatic plant that grows in irrigation ditches in the upper valley. Its leaves used as a soup condiment are boiled and eaten especially in times of food scarcity.

COMPOSITAE

Sonchus oleraceus L., *k'ana*, or sow thistle, is a major herbaceous weed in grain fields in the upper valley. Native to the Old World, it may have been introduced here in the sixteenth century along with wheat or barley. K'ana is fed to cattle.

Galinsoga parviflora Cav., *k'ora*, is a South American annual that is a very ubiquitous weed in maize fields in the upper valley. It too is fed to livestock.

Dahlia variabilis Desf., dahlia, is a cultivated ornamental that may have been brought to Peru from Mexico early in the colonial period. Although some kinds of dahlias have tuberous roots that have been consumed by Mexican Indians, there is no evidence that these are now nor ever have been eaten in the Andes. The dahlia is planted near dwellings and sometimes also in the corners of fields of other crops.

Lactuca sativa L., *lechuga* or lettuce, has been grown on a small scale since early colonial times. GARCILASO DE LA VEGA (1960:448) recalled eating lettuce in Yucay so it must have been an early introduction. Mestizos grow some lettuce but it is not popular among the peasantry.

Xanthium catharticum H.B.K., *allko kiska*, is an annual wild plant with spines usually found near human settlement. It has several uses in folk medicine: an infusion is made to treat ailments of the liver, spleen or kidneys; the leaves are also used externally to relieve rheumatism; and are also chewed for toothaches.

Cosmos peucedanifolius (Kuntze) Schaff., *panti*, is a red-flowering ornamental that grows wild in the upper valley. Blossoms of panti are often used during festive occasions such as weddings. They also are said to cure cough.

Bidens humilis H.B.K., *kiko*, is a common wild plant at higher altitudes in the valley, usually above 3,500 m. A yellow dye from the root was extracted before synthetic aniline dyes came into general use in the early twentieth century.

Ophryosporus origanoides H.B.K., *hetteka*, a branching shrub in the valley above 2,600 m, is used for firewood.

Polymnia sonchifolia Poepp. et Endl., *llacón*, is the only member of the Compositae to have been domesticated in South America. Llacón is a subtropical plant with edible sweet-tasting roots. The crop has been valued more as a sweet refreshment than as a nourishing food, for it is very watery and its food value is low. This plant is now cultivated sparingly in the Vilcanota Valley below 2,600 m and, in my opinion, is well on its way to extinction. In Inca times, there may have been a considerable movement of llacón to Cuzco for the festival of Inti Raymi which was an occasion for specialized food consumption and exchange.

Baccharis lanceolata Kunth, *chilka*, is one of the commonest native plants between 3,700 and 2,500 m and one of the most useful as well. This shrub, 1-2 meters high, rapidly and densely colonizes overgrazed pasture, abandoned crop land and cut-over eucalyptus groves. It is not a good forage, although goats will browse on it. Generally used to make baskets, the woody stems and branches are also burned as kindling. The ashes, which contain soda and potash, are used to make *llipta* when nothing better is available. It is also used in native medicine, although it may have been more important in COBO'S day (I:221) than at the present. The leaves are used in infusion against colds and are applied to the head as a poultice for headaches and wounds.

Baccharis odorata H.B.K., *tayanka*, is a very resinous shrub that is used as a combustible in the upper valley.

Senecio pseudotites Griseb., *maycha*, is a ruderal shrub common above 3,100 m. The leaves are applied in a poultice for bruises, rheumatism, and 'bad air'. In the early colonial period, it was used as a depurative (COBO: I:224-4). Folk healers in the valley still carry the name of *maycha médicos*, suggesting that this plant at one time may have been a panacea.

Mutisia viciaefolia Cav., *chinchirkuma*, is a common ruderal, about 1 meter high, that grows from 3,700 m to 2,800 m. The attractive orange flowers appeared in designs on Inca wooden cups and are today used ornamentally. The ashes of the stalks can be substituted for quinoa ashes to make llipta, used in extracting the alkaloid from coca leaves.

Barnadesia horrida L., *llaulli*, is a very thorny shrub often used as a living fence and as a fuel in village bakery ovens.

Tagetes minuta L., *wakatay* or Mexican marigold, is probably of Mesoamerican origin and was brought to the Andes in post-Conquest times (NEHER, 1968). *Wakatay* is the only New World condiment herb that is commonly used in the valley today. The leaves are also used in infusions and taken as a cure for intestinal gas, parasites and rheumatic pains. This plant appears to be spontaneous around peasant dwellings, but in many cases it is probably cultivated.

Tagetes mandonii Sch. Bip., *chikchipa*, may be a native herbaceous plant that was superseded by *T. minuta*. *Chikchipa* grows as a weed in corn fields and is collected for home use, but not seen in markets. It has use as a condiment and reportedly also in folk medicine. Peasants say that this plant is not 'fuerte' (strong) like *wakatay*, suggesting that it is lower in essential oils.

Matricaria chamomilla L., *manzanilla* or the German or wild camomile, is different from the common camomile (*Anthemis nobilis*) in genus but it has some similarities. Manzanilla is an herb that grows as a ruderal throughout the temperate zone of Europe and undoubtedly was brought to Peru by the Spaniards in early colonial times. Although it may be seen cultivated in the valley, it also volunteers vigorously around human settlements. In spite of its apparent ubiquitous presence, it is widely sold in native markets. The most valued part of the plant is the flower, although the stalk and the leaves are also used. An infusion of dried flowers is used for colic and stomach ache and an infusion of fresh flowers is indicated for sore throat, hemorrhoids, and eye and gum inflammation. It may also be drunk in place of tea or coffee as a refreshing hot beverage.

Grindelia boliviana Rusby, *chiri chiri*, is a yellow-flowering herb less than 1 meter high with extremely sticky buds and leaves. These leaves are collected from waste places above 1,500 m and used to cure aches and pains.

Eupatorium Sternbergianum D.C., *chamanway*, is a shrub, the leaves of which are used in an infusion against stomach ache. It is one of the best known medical plants in the upper valley.

Hypochoeris sonchiodes H.B.K., *miski pilli*, an herbaceous plant resembling chicory in appearance, grows along irrigation ditches in the upper valley. The leaves are boiled and eaten as a potherb.

Micania micrantha H.B.K., *wako*, is a rather rare woody vine with entire, leathery leaves which grows in the ceja, around 2,100 m. The juice of the leaves is used to treat snakebites, and the plant is very specifically associated with the remedy for the bite of the local poisonous species, the fer-de-lance (*víbora*).

4. SUMMARY AND CONCLUSIONS

The character and pattern of man's association with domesticated and wild plants in the Vilcanota Valley have been influenced primarily by the cultural context and by the physical limitations of the zone. The symbiotic relationship between these two kinds of living organisms has several dimensions of geographical interest which elicit summarization.

Plants and Man: The Historical Dimension

The pattern of man's use of the plant world is determined in part by the species available to him at different periods in his cultural development. Pre-agricultural, nomadic man had only the wild flora at his disposal; probably he was more thoroughly acquainted with its usefulness than most peasants are today. The plants which man has used over the longest period of time in the Vilcanota Valley are very likely members of the wild flora which today serve as famine foods and for medicinal purposes. At the same time the degree to which man depends on non-cultivated plants for food, shelter, fuel, ornamentals and medicinal needs has inevitably declined. A few serviceable species which were continuously exploited for many centuries are now rare, but the decline is best explained by the availability of more efficacious or accessible alternatives.

These alternatives became so successful that man long ago replaced much of the natural vegetation and native flora in the valley with domesticated plants. The current pattern of use in the valley is thus, in part, a distillation of the time-conditioned processes of plant introduction and genetic selection. Significant introductions from Mesoamerica and, since 1532, the Old World have together enriched the useful inventory by at least three times. With the sole but outstanding exception of the potato, the native Andean cultigens that today grow in the valley are pre-Inca relict holdovers that have not maintained their position in diet and economy. Yet since before the Inca period the same basic patterns of settled subsistence has continued and is still extant, for neither the social nor the technological aspects of modern industrial agriculture have as yet made a decisive impact on the valley.

Plants and Man: The Role of Perception

The particulars of plant use in the Vilcanota depression have been determined by man's past and present perception of each species' value to him. Although a few plants are used simply because they are thought to be useful, most of this perception has a rational basis. The important manifestations of plant use are related to the prevailing view of agriculture among the country people. Labor and time are matters of relatively low priority in the peasants' task of making a living from the land. Nor is the profit motive an overriding determinant, for instead of speculating on gaining from the uncertain, most farmers rely on those plants, varieties of plants, and techniques for working those plants which are time-tested, well understood, and virtually certain to supply the basic necessities. Such a conservative attitude may be a dedication to ancestral tradition for its own sake, but it also represents a means of grasping security which becomes a psychological requisite when people live at the very margins of existence. Assurance of a minimal food supply thus becomes highest priority in man's relationship to plants.

Plants and Man: The Ecological System

Most people today live in this valley zone as subsistence-oriented peasants. Their primary ecological adaptation is one of a direct transfer of energy from cultivated plants to man. Livestock, although fairly numerous, play a subsidiary role because animals are not ordinarily fed planted crop food and because fallowing of fields lowers dependency of the peasants on manure as fertilizer. Much of the valley landscape is covered with an artificially maintained, synthetic ecosystem we call agriculture. Dependence on non-cultivated, non-domesticated plants is normally low, but it periodically increases when agriculture fails to provide the expected energy.

An outstanding characteristic of man's association with plants in the Vilcanota Valley is the large extent to which the available vegetal inventory is known and utilized. This has resulted in practices noticeably different from those of modern industrial societies, such as polyculture, multiple functions of crop plants, ambiguous role of adventives, and pervasive awareness of the human-use possibility of all living plants. Survival is ordinarily assured by full and frugal use of the existing plant resources, but it is uncertain that maximization is necessarily a function of population density.

Plants and Man: The Spatial Patterns

The numerous physical variables from one end of the valley to the other account in large measure for the broad distributional patterns of plants and therefore of plant use. Microthermal, mesothermal, and macrothermal differences are undoubtedly the most important variables. Rainfall, soil, and landform also affect distributions. Compared to the physical variables, the cultural character of the population is distributionally uniform, and even if several distinct cultural groups inhabited the valley, it is likely that the small scale areal differentiation of plant use would still be primarily a function of this physical environmental diversity.

At a larger scale, however, cultural controls of plant distributions are evident within the limits imposed by physical factors. Clearing, irrigating, and terracing modify or eliminate certain aspects of physical control. The species and varieties of cultivated plants selected is another manifestation of this cultural control, and depend on a mixture of Old and New World traditions, and on the fact that the economy of most people is subsistence rather than commercial. Man determines the ultimate distribution of over three-fourths of the plant cover in the valley. Cultural controls may stretch absolute limits of the range of any single crop well beyond the limits one would expect on the basis of physical conditions. Human interventions in forming open habitats unwittingly determine the large-scale distributions of many elements of the useful plant inventory that are naturally dispersed.

Plants and Man: A Paradigm of Contemporary Culture Change

The Peruvian Andes, though remote from the mainstream of world commerce and ideas, presently show many signs of the global onslaught of Western technology and values. As modernizing influences of social, economic, and technological changes continue, one can expect a reorientation of the traditional relationship among plants, man and the land.

Increasing numbers of rural people will leave the land as they realize the rewards are not matched by their aspirations for a more comfortable material life. Marginal farming land, as on steep slopes, will cease to be cultivated, and a reasonable assumption is that homogeneous tree plantations will take over many of these sites. Wild plant use will become negligible and only certain individuals, rather than the masses, will be aware of the use potential of the flora.

The empirical approach to useful plants will in large part be replaced by advice, made on the basis of experiment, and proferred from the outside. Crop diversity will be replaced by relative gene uniformity. Native cultigens and local cultivars already are gradually being phased out by more efficient and/or better-tasting crop plants that were introduced from outside the Andean realm. When these and other changes occur, it is probable that the prevailing way of life as it is apparent today will be changed beyond recognition.

The Task Ahead

The nomothetic limitations of the cultural plant geography of a single valley are gross. Any thorough understanding of a non-literate people's perception and use of their environment needs a stronger empirical base than is currently available. The content and approach of this study is an initial step in developing conceptual strands on a geography of plant use, and needs to be complemented by additional empirical studies of plants and man in cultures not yet industrialized or at its incipient stages. Ultimately participants in primitive or folk cultures are unlikely to maintain their ways of life, even if they become fully aware of the environmental abuses and social disintegration of technologically superior groups. As an industrial, urban society descends on the Vilcanota Valley, one can expect a steady erosion of many traditional patterns of the plant-man relationship, and the folk character of the zone will be greatly modified by the turn of this century. Thus it becomes all the more imperative that interested scholars and institutions act now to record detailed information on the material culture of sub-industrial peoples.

BIBLIOGRAPHY

ACOSTA, J. 1962. Historia natural y moral de las Indias. Mexico: Fondo de Cultura Económica. 2nd ed. (1590)

AGRICULTURAL DEVELOPMENT OF PERU,THE. 1959. Rome: F.A.O. & I.B.R.D.

ALONSO DE HERRERA G. 1818. Agricultura general. Tomo I. Madrid: Imprenta Real. (1513)

ANALES DEL CUZCO: 1600-1750. 1901. Lima: Imprenta 'El Estado'.

ANDERSON E.. 1939. A classification of weeds and weed-like plants. *Science*. 89: 364-365.

ANONYMOUS MS n.d. Contiene varios puntos ... para el régimen buen gobierno de las haciendas de jesuitas sitas en el Cuzco y su jurisdicción. In: Biblioteca Nacional, Lima. (c. 1650-1700)

ANONYMOUS MS. 1649. (subject: haciendas in Calca). In: Biblioteca Nacional, Lima.

ANONYMOUS MS. 1729. Retasa del tributo que deben pagar los indios del repartimiento de Taray. Lima: Biblioteca nacional.

ANONYMOUS 1792. Descripción corográfica de la Provincia de Canes y Canches, conocida generalmente por el nombre de Tinta. *Mercurio peruano*. 5: 3-6; 9-14; 17-22..

ANONYMOUS. 1879. Descripción del Imperio. *Revista Peruana*. 401-427; 486-505.

ANONYMOUS. 1924. Libro de la visita general del virrey don Francisco de Toledo, 1570-1575. *Rev. Hist.* (Lima). 7: 115-216.

ANONYMOUS. 1957. Empadronamiento, medición y reparto de tierras de sembrar maíz, trigo y papa entre los indios de todos los ayllos de los pueblos del valle de Yucay, Guayllabamba y Urubamba ... *Rev. Arch. Hist. Cuzco*. 8: 389-432.

ANONYMOUS 1958. Repartición de tierras en 1658. *Rev. Arch. Hist. Cuzco*. 9: 426-434.

ANONYMOUS. 1963. Cartas, oficios y expedientes de los señores obispos del Cuzco. *Rev. Arch. Hist. Cuzco*. 11: 32-66.

BAUDIN, L. 1955. La vie quotidienne au temps des derniers Incas. Paris: Librarie Hachette.

BENNETT, W.C. 1946. The archeology of the Central Andes. In: Handbook of South American Indians. Vol. 2, Bull. 143, Bur. Amer. Ethnology. Washington.

BINGHAM, H. 1913. In the wonderland of Peru. *Nat. Geog. Mag.* 24 (4): 387-573.

BOTMILIAU. 1850. Les républiques de l'Amérique du Sud. La société péruvienne depuis l'indépendance. *Revue des Deux Mondes*. 6: 882-912.

BOWMAN, I. 1916. The Andes of Southern Peru. New York: Amer. Geogr. Soc. and Henry Holt & Co.

BROWN, W.L. 1960. Races of maize in the West Indies. Pub. 792, N.A.S. -N.R.C. Washington.

BUENO, C. 1951. Geografía del Perú virreinal. ed. D. VALCARCEL. Lima.

BUES C. 1917. Notas agrícolas de la Convención. *Bol. Min. Fomento* (Lima). 8: 5-51.

BUES C. 1935. La coca en el Perú. *Bol. Dir. Agric. y Gan.* (Lima). 5 (18): 3-72.

BUKASOV, S.M. 1930. The cultivated plants of Mexico, Guatemala and Colombia. MS Trans. at Mo. Bot. Garden from: *Bull. App. Bot. Gen. & Pl. Breed.* (Leningrad). Suppl. 47: 1-542.

CABALLERO FARFÁN, P. 1959. Supervivencias agrarias del Tahuantinsuyo. *Rev. Inst. Amer. Arte* (Cuzco). 9 (9): 121-152.

CÁRDENAS, M. 1950. Plantas alimenticias nativas de los Andes. *Folia Universitaria* (Cochabamba). 1-25.

CARRASCO, E. 1847. Calendario y guía de forasteros de la república peruana. Lima.

CÁSTELNAU F. DE. 1851. Expédition dans les parties centrales de l'Amérique du Sud. Vol. 4. Paris.
CÉSPEDES DEL CASTILLO, G. 1954. La renta del tobacco en el virreinato del Perú. *Rev. Hist.* (Lima). 21: 138-163.
CHAPMAN, F.M. 1921. The distribution of bird life in the Urubamba Valley of Peru. *Bull. U.S. Nat. Mus.* No. 117. Washington.
CHEESMAN E. 1944. Notes on the nomenclature, classification and possible relationships of cacao populations. *Trop. Agr.* (Trinidad). 21 (8) : 144-159.
CHIRIBOGA J. 1937. La malaria en los valles de La Convención. Lima: Dirección general de Salud Pública.
CIEZA DE LEÓN, P. 1883. The second part of the chronicle of Peru. (trans. C. Markham). London: Hakluyt Soc. Vol. 33. (1553)
CLARK, J.A. 1929. Distribution of the classes and varieties of wheat in the United States. *U.S.D.A. Bull.* No. 1498. Washington.
COBO B. 1956. Histora del nuevo mundo. Madrid: Ediciones Atlas. Vols. I & II. (1653)
COMPANIA CERVECERA DEL SUR DEL PERÚ LTA. 1962-63. Campaña agrícola del cultivo de cebada cervecera.
CONCOLORCORVO. 1959 (pseud. Alonso Cario de la Vandera). El lazarillo de ciegos caminantes. Madrid: Ediciones Atlas.
COOK O.F. 1916a. Agriculture and native vegetation in Peru. *Journ. Wash. Acad. Sci.* 6: 284-293.
COOK, O.F. 1916b. Quichua names of sweet potatoes. *Journ. Wash. Acad. Sci.* 6: 86-90.
COOK O.F. 1916c. Polynesian names of sweet potatoes. *Journ. Wash. Acad. Sci.* 6: 339-346.
COOK O.F. 1916d. Staircase farms of the ancients. *Nat. Geog. Mag.* 474-534.
COOK, O.F. 1919. Foot-plow agriculture in Peru. *Ann. Rept. Smith. Inst.* for 1918: 487-491.
COOK, O.F. 1925. Peru as a center of domestication. *Journ. Hered.* 16: 33-46; 95-110.
CORRELL, D.S. 1962. The potato and its wild relatives: section Tuberarium of the genus Solanum. Renner, Texas: Texas Research Foundation.
COUTU A.J. & R.A. KING. 1969. The agricultural development of Peru. New York: F.A. Praeger.
COVARRUBIAS POZO, J.M. 1961. Extracto y versión paleográfica del segundo libro del cabildo. *Rev. Univ. Cuzco.* 120: 299-372.
DARLINGTON, C.D. & P. WYLIE. 1956. Chromosome atlas of flowering plants. New York: Macmillan Co.
DEERR, N. 1928. Cane sugar. Manchester: Norman Rodger.
DELL, A. 1927. Llama land. New York: George H. Doran Co.
DICKINSON, J.C. 1969. The eucalypt in the Sierra of southern Peru. *Ann. Ass. Amer. Geog.*. 59: 294-307.
DOBYNS, H.F. 1966. Estimating aboriginal American population: an appraisal of techniques with a new hemispheric estimate. *Curr.Anthro.* 7 (4): 395-416.
EL EUCALYPTO EN LA REPOBLACIÓN FORESTAL. 1956. Rome: F.A.O.
ESTEVA FABREGAT, C. 1970. Un mercado en Chinchero. *Anuario Indigenista.* 30: 213-254.
FEJOS, P. 1944. Archaeological explorations in the Cordillera Vilcabamba, southeastern Peru. Viking Fund Pub. in Anthro. No.3 New York: Wenner-Gren Foundation.
FOSTER, G. 1960. Culture and conquest: America's Spanish heritage. Viking Fund Pub. in Anthro. No.27. New York: Wenner-Gren Foundation.
GADE D.W. 1966. Achira, the edible canna, its cultivation and use in the Peruvian Andes. *Econ. Bot.* 20 (4): 407-415.
GADE, D.W. 1967a. The guinea pig in Andean folk culture. *Geogr. Rev.* 57 (2): 213-224.
GADE, D.W. 1967b. Plant use and folk agriculture in the Vilcanota Valley of Peru: a cultural-historical geography of plant resources. Ph. D. dissertation, Department of Geography, University of Wisconsin (Madison). 442 pp.
GADE, D.W. 1968. Aranjuez of the New World. *Américas.* 20: 12-19.
GADE, D.W. 1969a. The llama, alpaca and vicuña: fact vs. fiction. *Journ Geogr.* 68 (6): 339-343.

GADE, D.W. 1969b. Vanishing crops of traditional agriculture: the case of tarwi (*Lupinus mutabilis*) in the Andes. *Proc. Ass. Am. Geog.* 1: 47-51.

GADE D.W. 1970a. The contributions of O.F. Cook to cultural geography. *Prof. Geog* 22 (4): 206-209.

GADE, D.W. 1970b. Coping with cosmic terror: the earthquake cult in Cuzco, Peru. *Am. Benedictine Rev.* 21 (2): 218-223.

GADE, D.W. 1970c. Ecología del robo agrícola en las tierras altas de los Andes centrales. *América Indígena*. 30 (1): 3-14.

GADE, D.W. 1970d. Ethnobotany of cañihua. (*Chenopodium pallidicaule*), rustic seed crop of the Altiplano. *Econ. Bot.* 24 (1): 55-61.

GADE,D.W. 1971. Grist milling with the horizontal water wheel in the Central Andes. *Tech. and Cult.* 12 (1): 53-61.

GADE D.W. 1972a. Bridge types in the Central Andes. *Ann. Ass. Am. Geog.* 62 (1): 94-109.

GADE, D.W. 1972b. Comercio y colonización en la zona de contacto entre la Sierra y las tierras bajas del valle del Urubamba, Perú. *Actas y memorias XXXIX Congreso internacional de americanistas* (Lima, Perú, 1970) 4: 207-221.

GADE D.W. 1972c. Red dye from Peruvian bugs. *Geog. Mag.* 45 (1): 58-62.

GADE D.W. 1972d. Setting the stage for domestication: brassica weeds in Andean peasant ecology. *Proc. Ass. Am. Geog.* 4: 38-40.

GADE, D.W. 1973. Environment and disease in the land use and settlement of Apurímac Department, Peru. *Geoforum*. no. 16: 37-45.

GADE, D.W. 1975. L'optique culturelle dans la géographie américaine. *Ann. Géogr.* (in press).

GADE D.W. & M. ESCOBAR. 1972. Canyons of the Apurímac. *Explorers Journal.* 51 (3): 135-140.

GADE, D.W. & R. RIOS. 1972. Chaquitaclla: the native footplough and its persistence in Central Andean agriculture. *Tools and Tillage*. 2 (1): 3-15.

GADEA, M. 1954. *Trigos españoles*. Madrid: Inst. Nac. de Invest. Agric.

GAGLIANO, J. 1963. The coca debate in colonial Peru. *The Americas*. 20 (1): 43-63.

GARCÍA RADA, G. 1936. Las royas en Arequipa y Cuzco. *Bol. Dir. Agric. Gan. y Colon.* 6 (20-21) : 17-50.

GARCILAZO DE LA VEGA, EL INCA. 1960. Comentarios reales. Cuzco: Univ. Nac. del Cuzco. (1609-1617)

GARNER H.F. 1959. Stratigraphic-sedimentary significance of contemporary climate and relief in four regions of the Andes mountains. *Bull. Geol. Soc. Am.* 70: 1327-1368.

GIRAL, F. 1966. Phyto-chemistry and zoo-chemistry in Hispanic America. *Impact of Science on Society*. 16 (4): 227-276.

GOHRING, R. 1934. La región del Cuzco. *Rev. Univ. Cuzco.* 23 (66): 19-40.

GOHRING, R. 1935. El Departamento del Cuzco. *Bol. Soc. Geog. Lima.* 52: 287-306.

GONZÁLEZ, E. 1833, Guía de forasteros del Departamento del Cuzco. Cuzco: Imprenta Pública.

GONZÁLEZ TAFUR, O.B. 1928. El cultivo de la caña de azucar y la elaboración de azúcar y alcohol en los valles de los Departamentos de Cuzco y Apuríimac. Informe No. 10 Soc. Nac. Agr. (Lima).

GRANDIDIER, E. 1861. Voyage dans l'Amérique du Sud: Pérou et Bolivie. Paris: Michel Lévy Frères.

GROBMAN, A., W. SALHUANA, R. SEVILLA & P. MANGELSDORF. 1961. Races of maize in Peru: their origins, evolution and classification. Pub. No. 915, N.A.S.-N.R.C., Washington.

GUAMAN POMA DE AYALA F. 1956/1966. La nueva crónica y buen gobierno (ed. L. BUSTIOS GALVEZ). Lima: Min. de Educación. 3 vols.

HAENKE,T. 1901. Descripción del Perú. Lima: Imprenta 'El Lucero.'

HARDY, O. 1914. Cuzco and Apurímac: a report on the commercial and industrial outlook of south central Peru. *Bull. Am. Geog. Soc.* 46 (7) : 500-512.

HARDY, O. 1919. The Indians of the Department of Cuzco. *Am. Anthr.* 21 (1): 1-27.

HARLAN, H. 1918. The identification of varieties of barley. *U.S.D.A. Bull.* 622. Washington.
HEISER JR., C.B. 1964. Chochos and other lupines in Ecuador. *Bull. Mo. Bot. Gar.* 52 (10): 8-13.
HEISER JR., C.B 1965. Cultivated plants and cultural diffusions in nuclear America. *Am. Anthr.* 67 (4): 930-949.
HEISER JR., C.B. & P.F. SMITH. 1953. The cultivated capsicum peppers. *Econ. Bot.* 7: 214-227.
HEMMING, J. 1970. The conquest of the Incas. New York: Harcourt Brace Jovanovich.
HERNDON, W.L. & L. GIBBON. 1854. Exploration of the Valley of the Amazon. Washington: A. Nicholson.
HERRERA, F.L. 1921. Contribución a la flora del Departamento del Cuzco. Cuzco: Imprenta 'El Trabajo.'
HERRERA, F.L. 1923. Plantas alimenticias indígenas cultivadas en el Departamento del Cuzco: maíz y sus variedades. *Bol. Soc. Geog. Lima* 40 (1-2) : 23-33.
HERRERA, F.L. 1929a. Formaciones vegetales del cañon del Urubamba. *Rev. Univ. Cuzco.* 13 : 159-167.
HERRERA, F.L. 1929b. Reseña histórica de los estudios botánicos verificados en el Departamento del Cuzco. *Rev. Univ. Cuzco.* 13 :205-223.
HERRERA, F.L. 1933. Filología quechua: botánica, etnológica: la duplicación de las voces en la nomenclatura indígena. *Rev. Mus. Nac.* (Lima). 2 (1) : 3-8.
HERRERA, F.L. 1941. Plantas alimenticias domesticadas par los Inkas cuyo uso se encuentra poco generalizado. *Chaski.* 1 (3) : 11-18.
HOBSBAWM, E.J.E. 1969. A case of neo-feudalism: La Convención, Peru. *Journ. Lat. Am. Stud.* 1 (1) : 31-50.
HODGE, W.H. 1949. Tuber foods of the old Incas. *Nat. Hist.* 58 (10) : 464-470.
HODGE, W.H. 1954. The edible arracacha. *Econ. Bot.* 8 : 195-221.
HORKHEIMER, H. 1960. Nahrung und Nahrungsgewinnung im vorspanischen Peru. Berlin: Colloquium Verlag.
HUMBOLDT, A. VON, 1811. Essais politiques, Paris.
JEFFREYS, M.D.W. 1963. The banana in the Americas. *Journ. Agric. Trop. Bot. Appl.* 10 (5-6-7) : 189-203.
JIMENEZ DE LA ESPADA, M. (ed.) 1956. Relaciones geográficas de las Indias. Madrid: Ediciones Atlas. (16th century)
JUAN Y SANTICILIA, J. & A. DE ULLOA. 1752. Voyages historiques dans l'Amérique méridionale. Amsterdam.
KOVALEVSKY, G.W. 1934. The vertical limits of crop plants in Peru. (in Russian). *Isvest. Gosud. Geog. Obetsch.* 66: 265-276.
KROEBER, A.L. 1939. Cultural and natural areas of native North America. (Univ. Calif. Pub. in Am. Arch Ethn. No. 38). Berkeley: Univ. Calif Press.
KUBLER, G. 1946. The Quechua in the colonial world. In: Handbook of South American Indians. Vol. 2, Bull. 143, Bur. Amer. Ethnology, Washington.
LA BARRE, W. 1947. Potato taxonomy among the Aymara Indians of Bolivia. *Acta Americana.* 5: 83-103.
LANNING, E.P. 1967. Peru before the Incas. Englewood Cliffs: Prentice-Hall Inc.
LATCHAM, R. 1936, La agricultura precolombiana en Chile y los paises vecinos. Santiago.
LATHROP, D.W. 1965. Origins of Central Andean civilization: new evidence. *Science.* 148 : 796-798.
LAVANDAIS, E. (pseud. Count de Sartigues). 1851. Voyage dans les républiques de l'Amérique du Sud. *Revue des Deux Mondes.* 10: 1019-1059.
LEON, J. 1964a. The maca a little known food plant of Peru. *Econ. Bot.* 18 (2): 122-127.
LEON, J. 1964b. Plantas alimenticias andinas. Boletín técnico No. 6. Instituto Interamericano de Ciencias Agrícolas, Lima.
LEVILLIER, R. 1925a. Ordenanzas dictadas por el virrey don Francisco de Toledo (1572). *Gobernantes del Perú.* 7 : 36-256.
LEVILLIER, R. 1925b. Ordenanzas del virrey don Francisco de Toledo (1575).*Gobernantes*

del Perú. 8 : 304-399.
LEWIN, B. (ed.). 1958. Descripción del virreinato del Perú. Rosario: Univ. Nac. del Litoral.
LOCKHART J. 1968. Spanish Peru 1532-1560: a colonial society. Madison: Univ. of Wisconsin Press.
LORENA, A. 1932. Las leguminosas arbóreas deYucay.*Rev. Univ. Cuzco.* 21 (63) : 99-103.
MACBRIDE J.F. 1936-. Flora of Peru. Field Musueum of Natural History, Botanical Series. Chicago.
MANGELSDORF, P., R. MACNEISH & W. GALINAT. 1964. Domestication of corn. *Science.* 143: 538-545.
MARCOY, P. 1875. Travels in South America. London: Blackie & Son.
MARÍN MORENO, F. 1961. Panorama fitogeográfico del Perú. *Rev. Univ. Cuzco.* 9-68.
MARKHAM, C.R. 1968. A history of Peru. New York: Greenwood Press.
MARTÍNEZ, H. Tierra y desarrollo de la comunidad en Canas y Canchis, Perú. *Journ. Inter-Amer. Stud.* 10 : 517-532.
MAURTUA V.M. 1906. Juicio de límites entre el Perú y Bolivia. Madrid: Imprenta Hijos de M.G. Hernandez. 12 vols.
MASON, J.A. 1957. The ancient civilizations of Peru. Baltimore: Penguin Books.
MAXWELL, T.J. 1956. Agricultural ceremonies in the Central Andes. *Ethnohistory.* 3 (1) : 46-71.
MEMORIA DE LA PROVINCIA DE URUBAMBA. 1874. *El Peruano.* 32 (97): 385.
MEJÍA XESSPE, M.T. 1931. Kausay: alimentación de los Indios. *Wirakocha.* 1 (1) : 9-24.
MENÉNDEZ RUA, A. 1948. Paso a la civilización. Quillabamba: Padres Dominicanos..
MIDDENDORF, E.W. 1895. Peru: Beobachtungen und Studien uber das Land und seine Bewohner: Heft III - Das Hochland von Peru. Berlin: Robert Oppenheim.
MINERVA DEL CUZCO. 1831. June 4. No.40.
MISHKIN, B. 1946. The contemporary Quechua. In: Handbook of South American Indians. Vol. 2, Bull. 143, Bur. Amer. Ethnology, Washington.
MONTESINOS, F. 1906. Anales del Perú. Madrid: Imp. de Gabriel L. y del Horno.
MOSCOSO CASTILLA, M. 1963. Secretos medicinales de la flora peruana y guía de la maternidad. Cuzco: Tip. Americana.
MURRA, J.V. 1960. Rite and crop in the Inca state. In: Culture in history. S. DIAMOND, ed. New York: Columbia Univ. Press. 393-407.
NEHER, R.T. 1968. The ethnobotany of *Tagetes. Econ. Bot.* 22 (4) : 317-325.
NICHOLSON, G.E. 1960. Chicha maize types and chicha manufacture in Peru. *Econ. Bot.* 14 (4): 290-300.
OCHOA, C. 1964. Recuentos cromosómicos y determinación sistemática de papas nativas cultivadas en el sur del Perú. *Anales Científicos* (La Molina). 2 (1) : 1-41.
OLARTE ESTRADA, J. 1967. Discontinuidad geográfico-física aguas abajo de Ollantaitambo. Geografía física y organizacíon del III Congreso nacional de geografía a Iquitos. Lima: Associación Nacional de Geografos. 92-102.
OSBORNE H. 1952. Indians of the Andes: Aymaras and Quechuas. London: Routledge & Kegan Paul.
PACHACUTI YAMQUI, J. de Santacruz. 1968. Crónicas peruanas de interés indígena. F. ESTEVE BARBA, ed. Madrid: Ediciones Atlas. 279-320.
PARDO, L.A. 1937. Ruinas precolombinas del Cuzco. Cuzco: Rojas.
PATIÑO, V.M. 1963. Plantas cultivadas y animales domésticos en América equinoccial. Cali: Imprenta Departamental. 2 vols.
PATIÑO, V.M. 1965. Historia de la actividad agropecuaria en América equinoccial. Cali: Imprenta Departamental.
PAZ SOLDÁN, F. 1877. Diccionario geográfico estadístico del Perú. Lima: Imprenta del Estado.
PAZ SOLDÁN M. & F. PAZ SOLDÁN. 1862. Geografía del Perú. Paris: Librería de Fermin Didot Herm.
PÉREZ VIDAL, J. 1954. La antigua técnica azucarera en el Perú. *Tradición* (Cuzco). 6 (15) : 91-95.

PERÚ. 1947. Censo nacional de población y ocupación de 1940. Lima: Ministerio de Hacienda y Comercio. Vol VIII (Cuzco).
PERÚ. 1966. Sexto censo nacional de población 1961. Lima: Dirección Nacional de Estadística y Censos. Vol. II (Cuzco).
PETERSON, R.F. 1965. Wheat. New York: Interscience Publishers.
PICKERSGILL, B. 1969. The archaeological record of chile peppers (*Capsicum* spp.) and the sequence of plant domestication in Peru. *Amer. Antiq.* 34 (1) :54-61.
PILARES POLO, V.M. 1936. Fitogeografía del valle del Cuzco. Cuzco: Imp. H.G. Rozas Suc.
PLAN REGIONAL PARA EL DESARROLLO DEL SUR DEI PERÚ. 1957. Informes. Lima.
POPONOE, W. 1922. Notes on Peruvian fruits. *Bull. Pan. Am. Union.* 55 (4) : 356-365.
POPONOE, W. & A. PACHANO. 1922. The capulin cherry. *Journ.Hered.* 13 : 50-62.
PUENTE Y OLLEA, M. 1900. Los trabajos geográficos de la Casa de Contratación. Seville: Esc. Tip. Y Libr. Salesianos.
PULGAR VIDAL, J. 1946. Historia y geografía del Perú: las ocho regiones naturales del Perú. Lima: Univ. Nac. de San Marcos.
QUEVEDO, C. 1900. Exploración de los indios antis del Cuzco, 1715. *Rev. Arch. y Biblio. Nac.* (Lima). 4 : 479-499.
RAIMONDI, A. 1898a. Itinerario de los viajes de Raimondi en el Perú: valle de Lares, Santa Ana y regreso por Mollepata y Limatambo. *Bol. Soc. Geog. Lima.* 8 : 1-61.
RAIMONDI, A. 1898b. Itinerario de los viajes de Raimondi en el Perú: Cuzco, Quispicanchis, Lucre, Pisac etc. hasta Marcapata. *Bol. Soc. Geog. Lima.* 8: 241-277.
REA J. & J. LEON. 1965. La mauka (*Mirabilis expansa* Ruiz & Pavon) un aporte de la agricultura andina prehispánica de Bolivia. *Anales Científicos* (La Molina). 3 (1) : 38-41.
RICK, C.M. 1950. *Capsicum pubescens*, a little known pungent pepper from Latin America. *Bull. Mo. Bot. Gar..* 38 (2) : 36-42.
ROBERTSON J.A. 1931. Some notes on the transfer by Spain of plants and animals to its colonies overseas. *The James Sprunt Hist. Stud.* 19 (2): 7-21.
RODRIGUES DEL RIVERO, A. 1947. Documento concernante a la introducción del trigo en América. *Rev. Arg. Agron.* 14 (2):173-175.
RODRIGUES SANDOVAL, L. 1949. Vida económica-social del indio libre de la sierra ecuatoriana. Washington: Cath. Univ. of America.
ROWE, J.H. 1946. Inca culture at the time of the Spanish Conquest. In: Handbook of South American Indians. Vol. 2, Bull. 143, Bur. Amer. Ethnology. Washington.
SABOGAL WIESSE J.R. 1965. El maíz en Chacán. *Anales Científicos* (La Molina) 3 (1): 68-88.
SAFFORD, W.E. 1916. Identity of cohoba, the narcotic stuff of ancient Haiti. *Journ. Wash. Acad. Sci.* 6:547-562.
SAUER, C.O. 1950. Cultivated plants of South and Central America. In: Handbook of South American Indians. Vol. 6, Bull. 143, Bur. Amer. Ethonology, Washington.
SAUER, C.O. 1952. Agricultural origins and dispersals. New York: Amer. Geog. Soc.
SAUER, C.O. 1965. Cultural factors in plant domestication in the New World. *Euphytica.* 14 (3):301-306.
SANTILLÁN, H. 1968. Relación del origen, descendencia, política y gobierno de los Incas. In: Cronicas peruanas de interés indígena. 7-149. Madrid: Ediciones Atlas.
SIMMONDS, N.W. 1959. Bananas. London: Longmans.
SQUIER, E.G. 1877. Peru: incidents of travel and exploration in the land of the Incas. New York: Harper & Bros.
STEIN, W.W. 1961. Hualcan: life in the Highlands of Peru. Ithaca: Cornell Univ. Press.
SWANSON, A. 1956. Alfalfa production in Peru. Special report No. 7. Programa Cooperativo de Experimentación Agropecuaria. La Molina.
TELLO, J. 1930. Andean civilization: some problems of Peruvian archaeology. *Proc. Int. Cong. Amer.* 1928. 259-290.
TESSMANN, G. 1930. Die Indianer Nordost-Perus. Hamburg: Friederichsen de Gruyter & Co.
TOSI, J.A. 1960. Zonas de vida natural en el Perú. Bol. No. 7, Zona andina, O.A.S., Lima.

TOWLE, M.A. 1961. The ethnobotany of pre-columbian Peru. Viking Fund Pub. in Anthro. No. 30. Chicago: Aldine Pub. Co.
TROLL, C. 1943. Die Stellung der Indianer Hochkulturen in Landschaftsaufbau der tropischen Anden. *Zeit Gesell. Erdkunde.* 3:93-128.
TROLL, C. 1968. The cordilleras of the tropical Americas. In: Geo-ecology of mountainous regions of the tropical Americas. Bd. 9 Colloquiium Geographicum. Bonn: Geographischen Institut. 15-55.
TSCHUDI, J.J. von. 1849. Travels in Peru during the years 1838-1842. New York: G.P. Putnam.
VALDEZ DE LA TORRE, C. 1921. Evolución de las communidades de indígenas. Lima: Ed. Euforion.
VALDIZÁN, H. & A. MALDONADO. 1922. La medicina popular peruana: contribución al folk-lore médico del Perú. Lima: Imp. Torres Aguirre.
VARGAS, C., C. 1936. El Solanum tuberosum a través del desenvolvimiento de las actividades humanas. *Rev. Mus. Nac.* (Lima). 5 (2):193-248.
VARGAS C., C. 1943. Ceja de montaña, una formación fitogeográfica de los Andes orientales del Perú. *Chronica Botanica.* 7:422.
VARGAS C., C. 1962. Phytomorphic representations of the ancient Peruvians. *Econ. Bot.* 16 (2): 106-115.
VAVILOV, N.I. 1949-50. The origin, immunity and breeding of cultivated plants. Waltham: Chronica Botanica Co.
VÁZQUEZ, M.C. 1967. La chicha en los países andinos. *América Indígena.* 27 (2):265-282.
VÁZQUEZ DE ESPINOSA, A. 1942. Compendium and description of the West Indies. (Smithsonian Miscell. Coll., vol. 102). Washington.
WACHTEL, N. 1971. La vision des vaincus: Les Indiens du Pérou devant la Conquête espagnole 1530-1570. Paris: Gallimard.
WEBERBAUER, A. 1945. El mundo vegetal de los Andes peruanos. Lima: Min. de Agricultura. (original German ed. 1911).
WEBSTER, S.S. 1970. The contemporary Quechua indigenous culture of Highland Peru: an annotated bibliography. *Behavior Sciences Notes* (New Haven). 5 (2):71-96; 5 (3):213-247.
WEST, R.C. 1948. Cultural geography of the modern Tarascan area. Smithsonian Institution. Inst. of Soc. Anth. Pub. No. 7. Washington.
WEINER, C. 1880. Pérou et Bolivie. Paris: Hachette.
WHITAKER, A.D. 1929. The Spanish contribution to American agriculture. *Agr. Hist.* 3 (1):1-14.
WHITAKER, T.W. & G.N.DAVIS.1962. Cucurbits: bontany, cultivation and utilization. New York: Interscience Pub.
WHITE, P.L. et al. 1955. Nutrient content and protein quality of quinoa and cañihua, edible seed products of the Andes mountains. *Journ. Agric. Food Chem.* 3 (6):531-534.
YACOVLEFF, E. & F.L. HERRERA. 1934-35. El mundo vegetal de los antiguos peruanos. *Rev. Mus. Nac.* (Lima). 3:241-322; 4:29-102.
YOUNG, R.A. 1924. Taros and yautias; promising new food plants for the South. *U.S.D.A. Bull.* 1247. Washington.
ZÁRATE, A. 1968. The discovery and conquest of Peru. (trans. J. Cohen). Baltimore: Penguin Books.
ZIMMERMAN, A.F. 1938. Francisco de Toledo, fifth viceroy of Peru, 1569-1681. Caldwell, Idaho: Caxton Printers.

CLOSSARY OF SELECTED TERMS

ACHIOTE – a tree yielding a seed dye
ACHIRA – a crop with edible rhizomes
ACHOKCHA – a native vegetable
aguardiente – rum
algarrobo – a tree resembling mesquite
allegado – one who sub-rents a parcel of land
Altiplano – a high plateau above 3800 m. in southern Peru and northern Bolivia, lying south of La Raya Pass
anden – a man-made agricultural terrace (often Inca)
AñU – an edible tuber crop
API – finely ground food
aporque – a cultivation practice which involves banking of earth and forming channels for irrigation water
ARRACACHA – a root crop
arrendire – one who rents land from an estate owner
arroba – a weight unit, equalling 25 pounds
ASNAPA – herb bundle
AYLLU – Indian community

bejuco – a liana or vine

calabaza – a cucurbit (*C. moschata*)
canchón – garden enclosure
capulí – a semi-domesticated cherry
carbón – a fungus disease of grain
carrizo – reed
cedro – a hardwood tree of the 'ceja'
ceja de la montaña ('ceja') – the densely forested area between 2400 m and 1550 m
CHACHACOMO – a wild tree of the upper valley
CHACRA – a) form or b) field on a farm
CHAKITAKLLA – the Andean foot-plow
CHAKMAY – plowing
CHALA – dried maize stalks and leaves
CHAMAN – a small, resinous tree
CHAMBA – a wild leguminous tree of the lower valley
CHANCACA – crude brown sugar
CHERIMOYA – an Andean domesticated fruit
CHICHA – a) (if term is not qualified) fermented maize beverage or b) fermented drink of another plant
CHILKA – a common shrub (*Baccharis*)
CHINCHIRKUMA – *Mutisia*, a flowering shrub
cholo – a partially Westernized person of Indian blood
CHULLPI – sweet corn
CHUNCHO – a) forest Indian or b) a hard maize cultivar of the lower valley
CHUñO – dehydrated potato

CHUPE – thick stew
chuzo – an adjective used to describe a harvested product which is shriveled and small

encomienda – the grant of Indian labor in the early colonial period
era – threshing floor

fanega – a unit of weight, generally equal to 25 pounds

guano de isla – bird excrement from coast, used as fertilizer

haba – a broad bean (*Vicia faba*)
hacendado – the owner of a manor estate
hacienda – manor estate
HATUN TARPUY – the major sowing of a particular crop
HATUPA – edible tumescence on maize
hortaliza – the general term for 'vegetable'
HULLKA – pitch fork

ICHU – a hardy grass (*Stipa ichu*)
ICHUNA – sickle
ILLA – potato worm
ingenio – sugar cane mill moved by animals

jora – sprouted maize

KANCHA – toasted maize
KAñIWA – a domesticated, hardy chenopod
KANTU – a red-flowering shrub
K'ASUNA – clod crusher
K'AYA – dehydrated oca
K'EPA TARPUY – the late planting of maize
KISHWAR – a useful tree (*Buddleia incana*)
KIWICHA – domesticated amaranth
KUTI – a tool used in cultivation ('lampa')
KUY – guinea pig

LAMBRAN – alder tree
lampa – kuti
LAWA – coarsely ground food
LAYMI – communally cultivated lands
LINGLI – dehydrated ullucu
LLACÓN – a domesticated root crop
LLANTÉN – *Plantago* herb
LLIPECHA – green chenopod leaves used as potherb
LLIPTA – a calcium ball of quinoa ash chewed with coca
LLOKE – a wild tree with valued wood
LLUNKA – peeled wheat
LUCMA – a native domesticated tree

MALLANWAY – a parasitic plant
manzanilla – camomile
MATEK'LO – an umbelliferous herb
MATO – the harvest from a perennial crop
MAWAY – early grown potatoes

mayordomo – estate manager
menestras – legumes used as food
mestizo – literally, 'mixed blood', as used here, the middle class
MISKA TARPUY – early sowing
MOKLLE – seed
MORAYA – high quality dehydrated potato
MOTE – boiled and peeled maize
MUñA – an aromatic native herb
MUTUY – *Cassia* shrub

nabo – *Brassica* weed (not the true turnip as used in valley)
ñUKCHU – an ornamental wild plant

OCA – a tuber crop

PAIKO – a ruderal chenopod used in medicine
palillo – a yellow dye substance
pampa – a flattish piece of land, large or small
PAPA – potato
PAPAKURU – a potato worm
pepino – a domesticated solanaceous fruit
PISONAY – an ornamental tree
PIKTU – wicker basket

quebrada – (local usage only) the valley depression above 2500 m
quinta – a small farm operated by a 'mestizo'
quintal – 4 arrobas or 100 pounds

RAUKANA – a small native digging tool
reducciones – the resettlement of Indians in the 16th century
ROKKE – a spiny but useful wild shrub
roya – *Puccinia* wheat rust

SAPALLU – a cucurbit (*C. maxima*)
SARA – maize
serrano – highlander
Sierra – the Highlands of Peru
SILLWI – a larval worm in maize and potato fields
SORA – pathological fungus (*Phytophthora infestans*)

TAKKE – two maize ears grown together
TAMBO – wayfarer's post
TARA – a hedgerow tree in the upper valley
temporal – an adjective describing an unirrigated field
terreno – field
TOPO – a variable unit area of land
TOROK – *Cecropia*, a weed tree of the lower valley
trapiche – pressing mill moved by a wheel powered by hand, animals or water
tree tomato – a fruit tree (*Cyphomandra*)

UCHU – capsicum pepper
ULLUCU – a tuber crop
UNCUCHA – an araceous root crop
UNKA – a wild tree valued for its wood

valle – as used in the Vilcanota Valley, that part of the depression below 1500 m

WAIKO – boiled food
WAKI – exchange by barter
WAKATAY – Mexican marigold (an herb)
WARANWAY – a tree (*Tecoma*) of the upper valley
WATIA – earthen oven for baking
WAYRURU – a brilliant red bean, used as a love charm
WILKA – *Piptadenia* tree of the lower valley
WIñAPO – sprouted maize
WIRO – green maize stalk for chewing
WITOK – *Genipa* tree and fruit

YUNCA (YUNGA) – the lower parts of the eastern valleys, generally below 1500 m
YUYO – potherb (sometimes restricted to wild mustard)

INDEX

GPSR Compliance

The European Union's (EU) General Product Safety Regulation (GPSR) is a set of rules that requires consumer products to be safe and our obligations to ensure this.

If you have any concerns about our products, you can contact us on ProductSafety@springernature.com

In case Publisher is established outside the EU, the EU authorized representative is:

Springer Nature Customer Service Center GmbH
Europaplatz 3
69115 Heidelberg, Germany

Batch number: 10607320

Printed by Printforce, the Netherlands